JUDAISM AND AI

Dignity and Moral Responsibility

BY
MOSHE PITCHON

Printed in the United States of America.
ISBN 979-8-9925712-7-1
21stCenturyJudaism.com publisher

This book is dedicated to Isidore and Marcel Philosophe.

I never discussed the arguments of this book with them directly. Yet whenever I sought to picture what an educated and committed Jew looks like—thinking about Israel, world Jewry, and the state of the world without ideology, but with openness and responsibility—they came to my mind.

From the first moment we met, and through the years of almost weekly lunches and occasional dinners, they offered something rare: steady encouragement without instruction, support without agreement, presence without agenda. They may not share all my conclusions, but they consistently reaffirmed my belief in what kind of human being Judaism informs when it is not reduced to ritual or sect.

Marcel and Isidore embody a Judaism engaged with the world, attentive to new technologies, committed to charity, and sustained by a living awareness of the covenant that binds the Jewish people across time and space.

For more than they likely know, I thank them.

Table of Contents

Preface

This book was written slowly.

Not because its subject demanded patience—artificial intelligence moves faster than thought—but because the questions it raises cannot be answered at speed. They concern responsibility, judgment, and the conditions under which a human being can still say *I answer for this*. Those conditions are fragile. They collapse when treated instrumentally.

I did not begin this book intending to write about technology. I began it with a concern that preceded AI: the quiet disappearance of answerability in modern life. Again, and again, I encountered situations in which harm occurred without anyone clearly responsible; decisions were made without decision-makers; power was exercised without a voice that could be addressed. The explanations were always plausible. The responsibility was always elsewhere.

Artificial intelligence did not create this condition. It revealed it.

What disturbed me was not the intelligence of machines, but the ease with which human beings were relieved of burden. Delegation felt like progress. Speed felt like necessity. Silence felt like prudence. Responsibility became something to be managed rather than assumed.

Judaism gave me the language to see what was happening—not as nostalgia, and not as theology, but as moral structure. Long before modern systems existed, Judaism understood that responsibility depends on delay, that power must be interruptible, and that silence in the face of authority is rarely innocent. It understood that law can restrain power, but cannot replace judgment; that obedience is never the end of obligation; and that even God, in the biblical and rabbinic imagination, remains bound by covenant.

These insights are not relics. They are diagnostic tools.

This book does not defend Judaism as an identity, a faith, or a community. It treats Judaism as a civilizational grammar—one that exposes what modern systems quietly undo. By placing this grammar alongside the history of science, industry, bureaucracy, and digital systems, the book asks a single question from multiple angles: *What happens to moral life when action no longer requires a responsible subject?*

This is a work of philosophy, not a political intervention. It does not proceed from a partisan agenda, nor does it seek to advance any contemporary ideological position. If its conclusions challenge existing arrangements of power, that is because philosophy—when taken seriously—tests claims of necessity and authority wherever they arise. Although I am a rabbi, I do not write from within any denomination, nor do I identify with any sectarian interpretation of Judaism. The rabbinic task, at its origin, was never to safeguard a particular lifestyle or closed system of interpretation, but to preserve Judaism's original moral intention: to protect responsibility against its displacement by power, habit, or certainty. This book should therefore be read neither as political advocacy nor as religious apologetics, but as an effort to remain faithful to responsibility itself.

I have resisted the temptation to offer solutions. Not because the problem is insoluble, but because responsibility is not something that can be engineered. It is a posture. A willingness. A refusal of refuge. Once responsibility is treated as a feature to be added, it has already been lost.

The appendices turn this critique inward. That was not optional. A moral argument that cannot be applied to one's own community is not an argument; it is a performance. Judaism itself now faces the temptations it once named: authority without accountability, power without interruption, silence justified as loyalty. To exempt Judaism from critique would betray the very tradition this book relies on.

This book is offered neither as warning nor as prophecy. It is an attempt to describe, as precisely as possible, the moral condition we are entering—and the cost of entering it without resistance.

Whether responsibility can still be recovered is not for this book to decide. Judaism never promised success. It demanded fidelity.

If this book has done its work, it will not leave the reader reassured. It will leave the reader addressed.

That, at least, was the responsibility I tried to honor in writing it.

Introduction

Responsibility After Intelligence

This is not a book about artificial intelligence.

At least, it is not a book about AI in the sense that now dominates public discussion: faster machines, smarter systems, existential risk, economic disruption, or regulatory gaps. Those questions matter, but they are secondary. They concern tools. This book concerns the human being who builds them—and the moral conditions under which responsibility can still exist.

The central claim of this book is simple and unsettling: the greatest danger posed by artificial intelligence is not that machines will become too powerful, but that human beings will cease to be responsible. Not because they choose irresponsibility, but because the structures through which action now occurs no longer require a responsible subject at all.

AI does not introduce this danger. It completes a long trajectory.

Modern civilization has been steadily reorganizing human action around speed, scale, systems, and optimization. Responsibility—once understood as answerability to others, to law, to God, or to history—has been thinned, deferred, and redistributed into processes too complex to interrupt and authorities too diffuse to address. Artificial intelligence accelerates this process to a breaking point. It does not merely act faster than human judgment; it acts in ways that dissolve the very conditions under which judgment could occur.

This is why familiar responses fail. Regulation governs tools, not meaning. Ethics presupposes agents who can pause. Transparency explains outcomes, but explanation is not responsibility. "Human-in-the-loop" models insert people into systems whose tempo renders intervention performative. Each solution addresses consequences while leaving the deeper failure intact.

That failure is anthropological.

Responsibility is not a value one adds to action after the fact. It is a structure of time, authority, and subjectivity that must exist *before* action occurs. It requires delay. It requires interruption. It requires a subject who can be addressed, questioned, and held to account. Where these conditions disappear, moral life does not evolve—it evaporates.

Judaism matters here not as theology, but as moral architecture.

Long before modern science, industrialization, bureaucracy, or digital systems, Judaism articulated a radically different understanding of what it means to be human in history. It rejected fate as moral explanation. It refused the idea that power is self-justifying. It insisted that reality itself must be intelligible enough to be judged. It trained human beings to live within obligation rather than inevitability.

Judaism is not defined by belief, ritual, or law alone. It is defined by *answerability*. Life speaks, and the human being must answer. Silence is not humility; silence does not remove one from involvement; it fixes one's involvement without examination. Obedience is never the end of responsibility. Even God, in the biblical and rabbinic imagination, is bound by covenant and subject to moral claim.

This book does not treat Judaism as a refuge from modernity. It treats it as a diagnostic instrument.

By tracing how responsibility was once structured—through time, law, memory, debate, and covenant—this book exposes what is now being lost. The argument proceeds historically, not nostalgically: from ancient fatalism to prophetic protest; from the Scientific Revolution's lawful universe to the Industrial Revolution's compression of time; from bureaucracy's diffusion of judgment to the digital turn's erosion of authorship; from AI's elimination of delay to the disappearance of the moral subject itself.

What emerges is not a critique of technology, but of civilization's willingness to live without being addressed.

This book is written against two comforting illusions.

The first is the belief that intelligence is the defining feature of the human being. Judaism never made this claim. Intelligence is instrumental; responsibility is constitutive. A being may calculate, predict, and optimize without understanding, judgment, or obligation. Artificial intelligence exposes this distinction with brutal clarity.

The second illusion is the belief that responsibility can be outsourced—to systems, institutions, experts, or procedures. Judaism denies this possibility. Responsibility is not transferable. It can be deferred, disguised, or denied—but never eliminated without hollowing out moral life itself.

The chapters that follow do not offer solutions, frameworks, or optimism. That is not an omission. It is fidelity to the problem. Responsibility cannot be engineered. It can only be assumed—or refused.

The appendices that conclude the book turn this argument inward, where it is most uncomfortable. They apply the same moral standard of responsibility to contemporary Jewish life, exposing how power, authority, and silence now threaten Judaism itself—from rabbinic monopolies that confuse law with the totality of Judaism, to political and communal failures of responsibility laid bare after October 7.

This book does not argue that artificial intelligence will destroy humanity.

It argues something more disturbing: that humanity may willingly abandon the burden that once defined it.

Whether that abandonment is inevitable is the final question this book leaves open. Judaism never promised inevitability. It promised obligation.

The question, then, is not what our systems will do.

It is whether we will still be there to answer when they act in our name.

PART I

Judaism Under Pressure

Chapter 1

Life as Address, Not Fate

J udaism begins with a refusal. A refusal to accept that reality is governed by fate, that power is self-justifying, or that meaning belongs exclusively to kings, gods, or natural forces. From the moment it made its appearance in history, Judaism insisted that existence is not mute. Life speaks. And because it speaks, it demands an answer.

This is the deepest rupture Judaism introduces into human consciousness: life is not something that merely happens to us. It addresses us. To live is to be summoned.

Ancient civilizations largely understood the world as governed by inevitability. The cosmos moved in cycles. The gods embodied forces. Human beings adapted, endured, and submitted. Meaning consisted in conforming to necessity, not in taking responsibility for it. Even when gods were worshipped, they were rarely questioned. Fate was tragic but unquestionable.

Judaism breaks this logic at its root.

The biblical perspective does not treat reality as fixed destiny but as moral address. The question it asks is not "What must happen?" but "What is being asked of me?" This shift relocates the human problem from survival within necessity to response within freedom.

The prophetic revolution begins precisely here.

The prophets of Israel do not oppose power as such. They oppose power without accountability. Their protest is not against kings, armies, or institutions per se, but against systems that claim inevitability—systems that demand obedience without justification.

The prophetic voice insists that nothing in heaven or on earth may operate without moral explanation. Not empire. Not law. Not even God.

This is why Abraham can ask, "*Will the Judge of all the earth not do justice?*"[1] This question is not heresy; it is remaining faithful to a binding commitment. God is not being denied. God is being held to account.

This insistence marks a fundamental change in how human beings understand their place in the world: Judaism does not merely speak of the divine, it demands moral coherence. Mystery no longer justifies authority; moral answerability takes its place.

From this moment on, history is no longer a theater of cycles. It becomes a field of responsibility.

To be Jewish is not primarily to affirm a proposition about God, but to stand in a relation of obligation. The human being is not asked to solve the mystery of existence, but to answer its moral claim.

This is why Judaism does not begin with metaphysics but with commandment. Not because commandment suppresses freedom, but because it defines freedom as answerability.

Life is not something we possess. It is something that addresses us.

This is the meaning of covenant ("*brit*"). Covenant is not contract. It is not agreement between equals. It is the binding of freedom to obligation. It assumes that human beings are capable of action, but not entitled to act without justification.

In mythic civilizations, fate explains suffering. In Judaism, fate is interrogated.

The biblical refusal of fate does not denies tragedy. It denies resignation. Suffering may occur, but it is never final in meaning. It always

---

1   *Genesis* 18:25

stands open to protest, repentance, and transformation.

This is why the Hebrew Bible preserves argument, dissent, and unresolved tension. The text itself models responsibility by refusing closure. Debate is not a failure of faith; it is its expression.

A world that cannot be argued with is not a moral world.

Judaism's core insight is that responsibility precedes power. The right to act is not earned by strength, intelligence, or success, but by answerability.

This principle governs everything that follows in Jewish thought.

The human being, in this account, is not a creature defined by capacity, intelligence, or productivity. The human being is defined by the ability to respond.

Judaism enters history by insisting that existence speaks — and that the human being is answerable.

Whether humanity will remember how to answer is the question that now lies before us.

Chapter 2

What Is a Human Being?

"What is the human being that You are mindful of him?"
— Psalm 8:4

Judaism approaches the question of the human being differently from almost every other civilization. It does not begin with biology, capacity, intelligence, or power. It begins with obligation.

The human being, in the Jewish conception, is not defined by what he can do, but by what is asked of him.

This distinction is decisive. Civilizations that define humanity by capacity inevitably rank human beings. Strength, intelligence, productivity, beauty, or usefulness become measures of worth. Those who possess more count for more. Those who possess less must justify their existence.

Judaism rejects this logic at its root.

The biblical claim that the human being is created *b'tzelem Elohim*—in *the image of God* [2]—has often been misunderstood as a metaphysical assertion about resemblance. But Judaism has never treated the image of God as a physical likeness or even as a set of superior faculties.

The image of God is not a capacity. It is a vocation.

---

2   *Genesis* 1:27

To bear the divine image is to be answerable. It is to stand in the world as a being whose actions matter beyond instinct, survival, or advantage. The image of God names not what the human being *is able* to do, but what the human being *is obligated* to do.

This is why Jewish thought never ties dignity to intelligence, creativity, or autonomy. Infants bear the image. The elderly bear it. The disabled bear it. Those who cannot speak, reason, or produce still stand fully within human dignity. Worth does not rise or fall with ability.

The image of God is not something that can be lost through weakness. It can only be betrayed through refusal of responsibility.

"You shall not stand idly by the blood of thy neighbor"[3]

Modern moral language is preoccupied with innocence: innocence from guilt, innocence from harm, innocence from complicity. Moral standing is often measured by distance—by how far one can stand from wrongdoing, power, or consequence. Judaism is deeply suspicious of this logic. It does not treat innocence as a moral achievement, nor does it promise exemption from the burdens of the world. To be human is not to stand outside history, causality, or obligation, but to be situated within them.

A world organized around innocence produces spectators rather than agents—those who watch, condemn, and withdraw, rather than act.

Judaism insists on a harder truth: moral dignity lies not in being free of involvement, but in accepting involvement without claiming purity. It is not innocence that sustains moral life, but the willingness to remain answerable when innocence is no longer available.

In the Hebrew Bible, the decisive moral failure is not transgression alone, but refusal. Cain's crime is not only murder; it is evasion. "*Am I my brother's keeper?*" [4] is the first attempt to escape responsibility by denying address.

---

3   *Leviticus* 19:16
4   *Genesis* 4:9

Judaism does not promise innocence. It demands response.

This is a difficult ethic. It leaves no refuge in purity, victimhood, or detachment. To be human is to be implicated. Life addresses us whether we want it to or not. Responsibility precedes consent.

This is why the biblical world contains no figure who can withdraw entirely from obligation. Even prophets resist their calling—but resistance itself confirms the claim. One can flee responsibility, but one cannot annul it.[5]

In modern thought, freedom is often defined as autonomy: the ability to choose without constraint. Judaism defines freedom differently.

Freedom is the capacity to respond-response-ability.

A being who can do whatever it wants is not free in the Jewish sense. A being who can answer—to others, to law, to God—is free. Constraint does not negate freedom; it gives freedom moral shape.

This is why commandment (*mitzvah*) is not the enemy of dignity but its condition. Commandment affirms that the human being is capable of responsibility and therefore worthy of being addressed.

The alternative—life without address— a world in which nothing calls, demands, or obligates-is not liberation. It is moral vacancy.

Judaism's anthropology is therefore fundamentally moral. The human being is a subject not because he thinks, but because he is accountable. Consciousness matters, but conscience matters more.

This distinction becomes critical under pressure.

Civilizations that ground human worth in intelligence eventually sacrifice the weak. Civilizations that ground worth in productivity even-

---

5   God has to ask Moses five times (*Exodus*. 3:11-4:20) before he accepts the task of leading the Israelites out of Egypt. Moses feels unworthy to accept such a huge responsibility, believes he does not speak well, thinks Pharaoh will pay no heed to him. "The influential Isaiah is one of the great biblical prophets to protest that he cannot be a prophet because of his "unclean lips: (*Isaiah*. 5: 5). Jeremiah objects because, like Moses, he does not speak well and is a mere boy (*Jeremiah*. 1: 6)."

tually discard the unproductive. Civilizations that ground worth in autonomy eventually abandon those who cannot choose.

Judaism grounds worth elsewhere: in the capacity to be bound by obligation—even when that obligation is burdensome, inconvenient, or painful.

The human being is the creature who can be commanded.

Every civilization carries within it a definition of the human being. That definition determines how it treats power, technology, law, and life itself. When the definition changes, everything changes with it.

If the human being is defined by intelligence, then superior intelligence becomes superior being.

If the human being is defined by performance, then failure becomes expendable.

If the human being is defined by autonomy, then dependency becomes shameful.

Judaism offers a different measure: answerability.

This measure has survived millennia because it does not compete with power. It restrains it. It does not glorify capacity. It governs it.

Before asking whether machines can think, a civilization must decide what thinking is for.

Before asking whether intelligence can be replicated, it must decide whether intelligence defines worth.

Judaism answers in advance: intelligence is not the source of dignity. Responsibility is.

If that answer is forgotten, no technological safeguard will matter.

If it is remembered, even unprecedented power can be governed.

The human being, in Judaism, is not the animal that thinks.

He is the being who must answer.

Chapter 3

Time, Delay, and Moral Formation

Judaism is a civilization built around time—not as a neutral medium through which events pass, but as the very condition in which moral responsibility becomes possible. Long before clocks measured efficiency and long before speed became a virtue, Judaism treated time as a moral structure.

Responsibility, in Jewish thought, is not instantaneous. It unfolds. It ripens. It requires duration.

No one is morally accountable for a reflex. Responsibility attaches only where there is space between impulse and action—space to consider, to hesitate, to imagine consequences, and to decide otherwise.

Judaism builds this space deliberately.

Law is not designed to produce rapid outcomes. It is designed to slow action enough for judgment to form. Testimony is examined. Evidence is contested. Decisions are debated. Minority opinions are preserved even when they do not prevail. The record itself becomes a moral memory of deliberation.

This slowness is not inefficiency. It is the structure that makes ethical judgment possible.

A civilization that moves too quickly cannot sustain responsibility, because responsibility requires the freedom not to act immediately.

Judaism trains human beings to wait.

This training appears everywhere. Blessings interrupt consumption. Dietary laws slow appetite. Prayer punctuates the day. Festivals impose rhythms that resist continuous productivity. Most radically, Shabbat halts all instrumental action once a week—not as rest alone, but as a declaration that the world does not belong to us to optimize endlessly.

By suspending creation, Judaism teaches that power must be voluntarily restrained. That the highest expression of freedom is not action, but the capacity to refrain. That the world is not exhausted by what can be done to it.

Delay, in this sense, is not passive. It is formative. It shapes a self capable of resisting immediacy.

Perhaps nowhere is Judaism's moral understanding of time is clearer than in *teshuvah*—repentance, return.

The concept of *teshuvah* arose from a basic human problem: moral failure is inevitable, but a moral world cannot survive if failure is final. A culture that demands responsibility must also provide a way for responsibility to continue after wrongdoing occurs. Without such a mechanism, moral life collapses either into despair or into denial.

If wrongdoing permanently defines the self, the individual either disengages ("there is no point in trying") or defends against responsibility through rationalization and blame-shifting. For that reason in the religion of ancient Israel, in distinction to its neighbors, ritual were not inherently efficacious. Judaism does not locate moral repair outside the human agent. There is no ritual shortcut that erases responsibility by transferring it elsewhere.

The language of *teshuvah* emerges directly from Scripture. Again and again, the prophets speak not of escape from judgment, but of return within it. *"Return to Me, and I will return to you."* [6] *"Let the wicked abandon his way... and return to the Lord."* [7] These are not appeals to

---

6   *Malachi* 3:7

7   *Isaiah* 55:7.

emotion but declarations about time: the future is not closed by the past.

Teshuvah is the refusal to let failure become fate. It stands against despair and against erasure alike.

The prophets do not soften judgment, but they do not allow it to close time. They insist that even after wrongdoing, the human being remains answerable—and therefore capable of return.

A world without *teshuvah* would be a world of verdicts without futures, of memory without repair. Judaism refuses such a world. It insists that responsibility does not end at the moment of transgression, but stretches forward, demanding acknowledgment, repair, and change. Return is not mercy alone; it is the condition under which moral life can endure.

Teshuvah presupposes that the future is not sealed by the past. That time remains open. That moral failure does not fix identity forever. Responsibility is not erased by error, but renewed through acknowledgment and change.

This is a radical claim. Judaism insists that moral agency persists because time itself can be reclaimed. *Teshuvah* is not psychological relief. It is a metaphysical statement: time can be re-entered responsibly. A civilization that loses this sense of temporal openness will eventually treat error as terminal—and responsibility as pointless.

Judaism preserves disagreement not as a flaw, but as a virtue. The Talmud famously records unresolved debates, minority positions, and rejected arguments alongside accepted rulings. Truth is not equated with unanimity.

"If all [the judges] find him guilty, he is acquitted."[8]

In capital cases, Jewish law required: A minimum court of 23 judges; a majority of at least two to convict. Never unanimity

---

8   Mishnah *Sanhedrin* 4: 1 expanded in the Babylonian Talmud, *Sanhedrin* 17a and 40a

If all judges voted guilty, the defendant was automatically acquitted.

The Talmud explains that: A unanimous guilty verdict indicates the court failed in its duty to search for exculpatory arguments. At least one judge must be willing to argue for acquittal, or the process is presumed prejudiced or rushed. Mercy and hesitation are not flaws but structural safeguards. As the Talmud puts it:

A court that sees only guilt has not judged fully[9]

This practice is impossible without time.

Debate requires patience. Listening requires restraint. Moral seriousness requires the willingness to remain with complexity rather than rushing toward closure. A culture that demands immediate resolution will sacrifice judgment for clarity.

Judaism prefers depth to speed.

The preservation of dissent is a way of keeping responsibility alive across generations. It signals that no decision exhausts meaning, and that future minds may be called to answer again.

Judaism distinguishes between time as experienced and time as controlled.

Ancient empires sought to master time through calendars, taxation cycles, and agricultural planning. Modern civilizations intensified this impulse through clocks, schedules, and deadlines. Time became something to be conquered, segmented, and exploited.

Judaism resists total mastery of time by sanctifying it.

Holy time interrupts instrumental time. It asserts that not all moments are equivalent and not all time belongs to human use. This resistance is not nostalgic. It is ethical.

A civilization that treats all time as available for optimization eventually treats human beings the same way.

---

9   *Sanhedrin* 17a

Moral responsibility cannot survive the elimination of delay.

Speed does not merely change how quickly decisions are made. It changes *who* makes them. When deliberation disappears, agency shifts. Judgment is replaced by reaction. Responsibility becomes performative.

Judaism understands this intuitively. That is why it builds pauses into life itself. That is why it refuses the fantasy of total immediacy. That is why it teaches that freedom is not the absence of limits, but the ability to live within them meaningfully.

Chapter 4

Power Without Answerability

J udaism is not suspicious of power because power is dangerous. It is suspicious of power because power lies.

Power always presents itself as necessity. It claims inevitability. It declares that things must be done "because there is no alternative." It speaks in the language of efficiency, security, order, and survival. And once power succeeds in presenting itself as unavoidable, it no longer needs to justify itself. Judaism enters history to deny power this exemption.

In the ancient world, power wore cosmic clothing. Kings ruled by divine mandate. Empires reflected the order of the heavens. Fate explained both suffering and hierarchy. What existed was what had to exist.

This was not merely political domination; it was metaphysical insulation. Power was protected from moral scrutiny because it was framed as natural, necessary, or sacred.

Judaism breaks this insulation.

The biblical understanding insists that power—any power—must answer. Kings are judged. Laws are questioned. Institutions are measured not by strength or stability but by justice. The prophets do not ask whether power works. They ask whether it is right.

This is unprecedented.

The prophets of Israel are not rebelling against authority as such. They are rebels against authority that refuses accountability. They speak to kings, priests, and people alike, and they speak in a single register: obligation.

"What gives you the right?"

"Who authorized this?"

"By what standard do you rule?"

These are not political questions. They are moral ones.

The prophets do not propose alternative empires. They refuse the premise that empire is self-justifying. Power must be explained. If it cannot be explained morally, it must be resisted—even if it is effective, popular, or stabilizing.

This stance is costly. It produces exile, persecution, marginality.

Jewish law is often misunderstood as obedience training. In fact, it is a system designed to restrain power.

Law slows decision-making. It demands procedure. It requires justification. It creates appeal. It preserves dissent. It binds even the powerful to standards they did not invent and cannot change at will.

Most importantly, law insists that no one—not king, priest, or prophet, President or Prime Minister—stands above judgment.

This is why the Torah is given publicly, not privately. Why it is written, debated, and transmitted through argument rather than decree. Law is not a tool of domination; it is a technology of accountability.

In this sense, Jewish law is not conservative. It is anti-tyrannical.

Power rarely announces itself as cruelty. More often, it announces itself as efficiency.

Things must be done quickly. Procedures must be simplified. Obstacles must be removed. Delays are irresponsible. Complexity is dangerous. The world is too urgent for hesitation.

Judaism recognizes this language. It has heard it before.

Efficiency is not neutral. It is a moral claim masquerading as practicality. When efficiency becomes the highest value, judgment is treated as friction and conscience as latency.

Judaism refuses this hierarchy.

The question is never "Does it work?" but "What does it do to the human being who wields it—and to the one subjected to it?"

In many civilizations, authority demands silence. Obedience is framed as virtue. Dissent is treated as disorder. Questioning is equated with disloyalty.

Judaism does the opposite.

It preserves argument. It sanctifies questioning. It records dissent. It allows—even demands—that authority be spoken to. Within such a tradition, silence is not neutrality; it functions as a refusal to bear responsibility.

This is why the biblical text itself is argumentative. Why God is questioned—by Abraham over the fate of Sodom, by Jeremiah and Job in the face of suffering, by Jonah who protests God's mercy as a betrayal of justice, and by later figures such as Rabbi Levi Yitshak of Berdichev, who famously interrupted the Yom Kippur service to accuse the King of heaven of failing to protect His people. Why outcomes are debated. Why no voice is final simply because it is powerful.

A world in which power cannot be questioned is not a moral world.

Judaism's insistence on answerability has never been rewarded with dominance. It has been punished with vulnerability.

Empires tolerate efficiency. They do not tolerate accountability.

This is why Judaism is repeatedly expelled, marginalized, or scapegoated. A people that refuses to accept inevitability threatens every system that depends on it. To insist that power must justify itself is to undermine the metaphysical foundations of domination.

Judaism survives not because it defeats power, but because it refuses to worship it.

Power without answerability is always dangerous—no matter how intelligent, well-intentioned, or efficient it appears.

When power accelerates beyond the speed of judgment, when authority hides behind complexity, when systems operate without explanation, responsibility erodes.

Judaism stands in history as a permanent refusal of that erosion.

Chapter 5

History as Moral Drama

J udaism does not understand history as a sequence of events. It understands history as a trial.

This is one of its most radical claims. In most ancient civilizations, history was cyclical, tragic, or opaque. Events repeated. Empires rose and fell. Suffering occurred, but meaning did not accumulate. What happened had no obligation attached to it beyond endurance.

Judaism breaks this frame.

History, in Jewish thought, is not merely what happens to humanity. It is what humanity is answerable for. Time is not a loop. It is a direction. And direction implies responsibility.

Pre-biblical cultures largely understood time as circular. Seasons returned. Kingdoms followed patterns. Fate ruled. Even when gods intervened, they did so within an unchanging cosmic order.

Judaism introduces something unprecedented: narrative time.

Narrative time has a past that matters, a present that demands response, and a future that remains open. Events are not interchangeable. What happens now alters what can happen later. Action leaves a moral residue.

This is why memory is sacred in Judaism. Remembering is not nostalgia; it is accountability. To remember is to accept that the past continues to claim the present.

"Remember that you were strangers in the land of Egypt"[10] is not a historical observation. It is a moral instruction. History addresses the living.

Judaism's linear understanding of history emerges from covenant.

Covenant (*b'rit*) is one of the organizing ideas of Judaism—no less fundamental than teshuvah or responsibility. It is not merely a theological doctrine but a comprehensive way of understanding human existence, community, and politics. The Hebrew Bible does not present covenant as a mystical bond or an abstract belief, but as a concrete, morally binding relationship that structures reality itself.

At its core, covenant is a relationship initiated and sustained through mutual commitment. It binds parties—God and Israel, Israelites with one another—through promises of fidelity that are voluntarily entered into and continuously renewed.

Covenant is therefore neither fate nor contract in the modern sense. One may belong to the community by destiny, but one enters the covenant by will. Judaism insists on this distinction: obligation is inherited, but responsibility is assumed.

For this reason, covenant is less a theological idea than a theo-political one. It functions as a form of social and political organization, comparable in importance to natural law in defining justice or to social contract theories in explaining political society.

Yet unlike modern contracts, covenant does not reduce obligation to self-interest, nor does it dissolve community into individual transactions. It establishes a partnership of equals—diverse yet bound—who commit themselves to shared purposes under a transcendent moral authority.

Covenant creates a people not through hierarchy or organic unity, but through consent and obligation. It allows human beings to remain free

---

10   *Deuteronomy* 24:22. The Torah repeats this command multiple times with slight variations (e.g., *Exodus* 22:20; *Leviticus* 19:34; *Deuteronomy* 10:19.

while being bound, to preserve individual integrity while sharing in a collective destiny. In contemporary terms, it is a federal relationship: unity without uniformity, diversity without fragmentation. This is why covenantal thinking has shaped Jewish understandings of politics, law, and civil society for millennia.

Crucially, covenant is never purely vertical. It does not link isolated individuals to God alone. It also establishes horizontal bonds of mutual responsibility. *"All Jews are responsible for one another"* [11] is not a moral slogan but a covenantal principle: the subject of the covenant is the collective people of Israel, across time. Covenant therefore means covenant-with-all-other-Jews, past and present. Responsibility is shared, inherited, and inescapable.

This structure transforms both God and the human being. In the covenantal conception, God is not an absolute monarch but a constitutional one, bound by promises just as Israel is bound by obligations. Divine power is constrained by law; authority submits to fidelity. At the same time, human freedom is acknowledged and institutionalized. God recognizes human agency and demands that humans accept the consequences of that freedom: to build lawful communities, pursue justice, enforce norms, and answer for failure.

The radical character of covenant lies here. Religion is lifted from magic to ethics. Power is subordinated to obligation. History becomes a shared project rather than a closed fate. Covenant insists that meaning arises not from domination or inevitability, but from partnership and accountability.

At its highest level, covenant is a community of souls—a bond of loyalty and care in which each member is responsible for all others. It is the political analogue of the I–Thou relationship in personal life: a way of being with others that preserves dignity, demands response, and refuses indifference. Silence, withdrawal, or neutrality are never covenantal stances. To be in covenant is to be addressed—and to answer.

---

11   Shevuot 39a; Sifra Lev. 26: 37

Finally, covenant is not static. Jewish history repeatedly reinterprets and re-articulates it under conditions of crisis. Exile, persecution, emancipation, and modernity do not abolish the covenant; they test and transform it. The covenant endures precisely because it is a living relationship, renewed through responsibility rather than guaranteed by power.

To live in covenant, then, is to inhabit a world in which freedom and obligation are inseparable, where community is chosen as well as inherited, and where responsibility—personal, collective, and historical—is the price of dignity.

Covenant binds action to consequence across time. It asserts that what human beings do matters not only immediately, but generationally.

Blessing and curse are not rewards and punishments in a mechanical sense. They are the moral shape of history itself. A society that acts justly becomes capable of sustaining life. A society that abandons responsibility corrodes its own future.

This is not determinism. It is moral causality.

Judaism refuses both fatalism and utopia. The future is neither fixed nor guaranteed. It is contested. It must be shaped through action.

The prophets and historical meaning

The prophets do not predict the future. They interpret the present.

This distinction is crucial. Prophecy in Judaism is not foresight; it is moral diagnosis. The prophet looks at injustice now and declares where it leads. History is read as consequence, not as spectacle.

This is why prophecy is uncomfortable. It refuses to allow power to hide behind success. It insists that prosperity can coexist with moral failure—and that such prosperity is fragile.

The prophet speaks because history can still change. If the future were sealed, speech would be pointless.

Judaism's most misunderstood idea is messianic hope.

Messianism is not the belief that history will end in perfection. It is the belief that history is unfinished. That redemption is possible but not inevitable. That the world is not yet what it must become.

This openness is ethical, not mystical. It means that no moment exhausts responsibility. No injustice is final. No failure forecloses the future completely.

Other traditions seek salvation outside history or enlightenment beyond it. Judaism insists that meaning must be realized within history, through action, law, and responsibility.

Judaism waits—not for history to end, but for it to be repaired.

Once history is understood as morally directed, it becomes a source of pressure. Every generation inherits unfinished obligations. The past demands reckoning. The future demands preparation.

This is why Judaism never allows itself to be complete. Its story has no final chapter. Its law continues to be interpreted. Its debates remain open.

Completion would mean closure. Closure would mean exemption from responsibility.

Judaism refuses that exemption.

What a civilization believes about history determines how it treats power, time, and the human being. If history is meaningless, responsibility dissolves. If history is cyclical, responsibility stagnates. If history is morally directed, responsibility deepens.

Judaism stands alone among ancient civilizations in insisting that the future is ethically shaped—and that human beings are accountable for shaping it.

This claim will now be tested under conditions no previous generation has faced.

For the first time, humanity is creating forces that act faster than historical judgment, systems that operate beyond memory, and powers that may no longer recognize obligation.

If history ceases to be a moral drama, responsibility will not merely weaken—it will vanish.

The question that now confronts us is not whether history will continue. It is whether it will still mean something.

A civilization rarely notices the moment it crosses a threshold. Change usually announces itself as progress: faster, stronger, more efficient, more capable. Only later does it become clear that something deeper has shifted—that the conditions under which responsibility once formed have quietly altered.

Judaism has always been attentive to this danger. Not because it fears power, but because it understands how power behaves when it escapes moral constraint.

The story told so far in this book was not technological. It was anthropological. It described a human being defined not by capacity but by answerability; a civilization organized around delay, law, debate, memory, and obligation; a conception of history as morally directed rather than mechanically repetitive. This architecture allowed Judaism to survive power without worshipping it.

But architecture alone does not stop pressure.

Every major technological transformation places stress on the moral structures that preceded it. When tools change what human beings can do, they also change what human beings are tempted to excuse. Speed begins to look like necessity. Sheer size begins to look like inevitability. Complexity begins to look like exemption.

At that point, power no longer needs to justify itself. It simply operates.

What follows now is not a history of inventions, but a history of thresholds—moments when humanity gained new powers faster than it learned how to govern them.

The scientific revolution did not merely produce knowledge; it redefined what counted as knowledge.

The industrial revolution did not merely amplify labor; it reordered time, space, and authority.

Each transformation expanded human reach while quietly eroding older forms of moral friction.

And yet, through all of this, something remained intact.

Despite telescopes and engines, despite railroads and telegraphs, despite factories and bureaucracies, the human being remained the final site of judgment. Machines extended bodies and senses. Systems accelerated action. But interpretation, meaning, and responsibility still belonged to human agents.

That condition is about to change.

Before confronting a technology that simulates judgment itself, it is necessary to understand how earlier revolutions prepared the ground. Not by intention, but by accumulation. Not by malice, but by momentum.

The part of this book that begins here traces these earlier thresholds—not to assign blame, but to clarify inheritance. The world that produces artificial intelligence did not appear suddenly. It was built, layer by layer, by civilizations that learned how to command nature long before they learned how to restrain themselves.

Judaism has walked through every one of these thresholds before. Sometimes as participant. Sometimes as critic. Often as conscience. Never as master.

The question now is whether that historical memory still matters—when power no longer arrives wearing a human face, and when decision itself threatens to detach from responsibility.

What follows is the story of how humanity learned to accelerate faster than it learned to answer.

And why that acceleration now confronts its final test.

PART II

Civilizational Thresholds

Chapter 6

The Scientific Revolution: When the World Became Legible

B etween the sixteenth and seventeenth centuries, knowledge ceased to be grounded primarily in inherited wisdom, sacred texts, or moral purpose, and came to be grounded instead in measurement, experiment, and predictive power.

This fundamental reorientation of how reality was understood what is commonly called the Scientific Revolution.

It did not begin as a revolt against God. It began as a revolt against opacity.

For most of human history, reality ruled through mystery. The cosmos was governed by forces that could not be questioned, only appeased. Storms arrived because gods willed them. Disease struck because fate demanded it. Power justified itself by appealing to the inscrutable. What could not be explained could not be challenged.

Judaism had already broken with this logic long before modern science emerged. The prophetic tradition rejected the idea that reality is mute or arbitrary. It demanded that existence answer to moral reason. *"Will the Judge of all the earth not do justice?"*[12] is not merely a theological question; it asserts that reality itself must be open to judgment. Reality must be intelligible enough to be judged.

---

12   *Genesis* 18:25

Making the world intelligible does not, by itself, make human beings responsible.

Knowledge expands power faster than it expands judgment. When explanation replaces obligation, mastery becomes morally weightless.

The Scientific Revolution inherits this demand—but strips it of covenant.

In pre-modern societies, mystery functioned as a form of rule. The less intelligible the world appeared, the more authority belonged to those who claimed privileged access to its secrets—priests, kings, oracles. Knowledge was not cumulative; it was custodial. Truth did not advance. It was guarded.

This was not ignorance. It was structure.

Opacity stabilized hierarchy. If the world could not be understood, it could not be questioned. If events were the result of divine whim or cosmic necessity, responsibility dissolved into submission.

Judaism had already protested this arrangement. The prophets refused to accept that suffering, injustice, or power could hide behind mystery. But prophecy alone could not reorganize the material understanding of the world.

Science would.

The Scientific Revolution begins when thinkers like Galileo, Kepler, and Newton insist on something radically simple: the universe operates according to laws that can be discovered, articulated, and tested. Reality does not merely occur. It obeys.

This is not a rejection of transcendence. It is a refusal to accept arbitrariness.

Nature becomes legible. The heavens are no longer the domain of divine caprice but of mathematical order. Motion follows rules. Forces can be measured. Causes can be traced. Effects can be predicted.

This is an ethical transformation disguised as a technical one.

To say that the world must make sense is to say that it cannot rule by silence. To demand explanation is to reject domination by mystery. In this sense, science is a secular continuation of the prophetic protest against opacity.

Yet something crucial changes.

The prophetic demand for intelligibility was always bound to responsibility. To understand the world was to be accountable for how one acted within it. Knowledge was not neutral. It obligated.

The Scientific Revolution severs this bond.

Science insists that reality be intelligible—but it does not insist that power be answerable. Nature can now be mastered without being morally interpreted. Knowledge accumulates faster than judgment. Explanation replaces obligation.

This is not a failure of science. It is a limit of its scope.

Science can tell us how the world works. It cannot tell us what the world asks of us.

Modern science redefines the human being as an observer. Detached. Neutral. Standing outside the system being studied. The knower is separated from the known.

This stance is immensely powerful. It allows for objectivity, replication, and control. But it also introduces distance—moral as well as epistemic.

When the world is treated as object, it becomes easier to manipulate without reflection. When the knower is detached, responsibility can be deferred.

Judaism had never allowed this separation. The human being was always inside the world, addressed by it, bound by obligation. Knowledge was inseparable from conduct.

The scientific stance alters this balance.

And yet—despite its revolutionary force—the Scientific Revolution leaves one crucial structure intact.

Human judgment remains central.

Machines do not decide. Instruments do not interpret meaning. Data does not govern action. The human mind still stands as the final arbiter of significance and responsibility.

Science extends vision. It does not replace conscience.

This distinction matters.

Because as long as interpretation remains human, responsibility can still be located. Power may increase, but answerability has not yet dissolved.

The Scientific Revolution liberates humanity from superstition and arbitrary authority. It makes the world intelligible. It empowers intervention. It saves lives.

But it also introduces a temptation: to believe that explanation is sufficient, that control is justification, that understanding replaces obligation.

Judaism resists this temptation.

To understand is not to be absolved. To explain is not to be excused. The world becoming legible does not mean it becomes morally neutral.

But at this stage, something still holds.

Human beings may now command nature—but they still command themselves. Judgment remains slow. Responsibility remains possible. Delay still exists between knowledge and action.

That delay will not last forever.

The Scientific Revolution opens the door to unprecedented power. It does not yet ask whether humanity is prepared to carry it.

That question will soon become unavoidable.

Chapter 7

The Industrial Revolution:
When Time Was Conquered

I f the Scientific Revolution made the world intelligible, the Industrial Revolution made it obedient.

For the first time in human history, power no longer waited for nature. Energy could be generated on demand, sustained indefinitely, and directed with precision. Rivers no longer determined where factories could stand. Seasons no longer governed when work could occur. Muscle—human or animal—was no longer the primary measure of production.

What changed was not merely how much humanity could do, but **when** it could do it.

Time itself was captured.

Prior to industrialization, power was intermittent. Wind rose and fell. Water flowed or froze. Harvests arrived when they arrived. Even empires were tethered to ecological rhythms they could not override.

These limits enforced delay.

They created pauses between intention and execution, between desire and fulfillment. Moral responsibility did not arise automatically from these pauses, but the pauses made responsibility possible. Power could not accelerate beyond human judgment because it could not accelerate beyond nature.

This constraint disappears with steam.

The steam engine is not revolutionary because it is powerful. It is revolutionary because it is *reliable.*

Unlike wind or water, steam does not wait. It responds. It can be summoned at will, sustained continuously, and intensified at scale. For the first time, energy becomes abstract—separable from environment, location, and season.

Power is no longer borrowed from nature. It is manufactured.

Responsibility depends on delay. When time becomes controllable, judgment becomes optional.

This marks a profound shift in the human relationship to time. Work no longer follows daylight. Production no longer respects exhaustion. Output no longer pauses because bodies do.

The machine does not tire. And increasingly, neither should the human.

Industrialization reorganizes life around the clock.

Time becomes divisible, quantifiable, and exchangeable. Hours can be bought and sold. Productivity can be measured. Efficiency becomes the highest virtue.

This transformation is not neutral. It reshapes the human being.

Under clock-time, delay appears as waste. Reflection becomes inefficiency. Rest becomes suspect. Judgment is pressured to keep pace with output. The slower rhythms through which moral deliberation once unfolded are treated as obstacles to progress.

The human being is no longer the measure of time. Time becomes the measure of the human.

Factories do not merely produce goods. They produce discipline.

Bodies are synchronized. Movements standardized. Tasks fragmented. Decision-making is centralized. Responsibility is diffused across systems no single individual fully controls.

This diffusion matters.

When action is broken into parts, accountability becomes difficult to locate. No one person causes harm. Everyone follows procedure. The system moves, and individuals move with it.

Power expands—but answerability thins.

This is not cruelty by intent. It is harm by structure.

Industrial society advances faster than its moral frameworks can adapt.

New forms of exploitation emerge before new forms of protection can be articulated. Law struggles to catch up with scale. Ethics lags behind capacity. The speed of production outpaces the speed of judgment.

Judaism recognizes this pattern.

It has always understood that when power accelerates, responsibility must be reinforced deliberately—or it will be overwhelmed. The industrial age does not destroy responsibility, but it places it under unprecedented strain.

And yet, even here, something crucial remains intact.

Machines produce. Humans decide.

Despite mechanization, judgment is still exercised by persons. Orders are given by human authorities. Goals are set by human institutions. Meaning is still interpreted by human minds.

The steam engine extends force. It does not simulate judgment.

This distinction preserves a final moral anchor. As long as decisions remain human, responsibility can still be assigned—even if imperfectly, even if unjustly.

The center holds, though it trembles.

The great temptation of the Industrial Revolution is to mistake speed for necessity.

When systems move quickly, slowing them appears irresponsible. When productivity rises, restraint looks irrational. When efficiency delivers results, deliberation feels indulgent.

This temptation does not erase conscience. It pressures it.

Judaism resists by insisting that power never justifies itself. That speed does not absolve. That systems do not replace judgment. That responsibility cannot be delegated to machinery.

But resistance becomes harder as scale increases.

This chapter marks the second modern threshold: time becomes controllable.

For now.

The Industrial Revolution conquers time—but it does not yet conquer judgment. The human being still stands between power and action.

That position will soon become unstable.

Chapter 8

The Collapse of Distance: Train, Telegraph, and the Compression of the World.

T he Industrial Revolution conquered time. The next revolution collapsed distance.

Until the nineteenth century, power remained tied to physical presence. To rule was to be near. To command was to stand within sight, sound, or reach. Authority required embodiment. Decisions traveled at the speed of bodies—on foot, on horseback, by ship. Distance enforced delay, and delay preserved a measure of moral friction. Moral responsibility weakens when action no longer requires encounter.

That friction was about to disappear.

The railway did more than move people faster. It reorganized how human beings experienced space.

Before trains, distance was physical and experiential. A city was far because the journey was arduous. Separation was measured in fatigue, danger, and time away from home. With the railway, distance shrank not in miles, but in meaning.

Cities moved closer without moving at all.

Suddenly, what had required days required hours. Regions that had felt remote became adjacent. The map did not change—but the human sense of the world did. Space became psychological rather than geographical.

This transformation forced a new discipline upon time itself. Local time was no longer sufficient. Railways demanded synchronization. Clocks were standardized. The world began to run on shared minutes rather than lived rhythms.

Time was no longer personal. It was systemic.

The deeper shift was not speed, but **presence without proximity.**

The train allowed power to act at a distance while retaining coherence. Orders could be issued centrally and enacted far away without the ruler ever arriving. Economic decisions could reshape communities the decision-maker would never see.

Action no longer required encounter.

This mattered morally. Presence had always imposed a minimal responsibility. To see the effects of one's decisions was to be affected by them. Distance had functioned as a weak but real ethical brake.

The railway weakened that brake.

If the train collapsed space, the telegraph annihilated waiting.

For the first time in human history, a message could arrive before its sender. Thought detached from movement. Words outran bodies. Command outran presence entirely.

This was not merely faster communication. It was a new condition of action.

Decisions could now be made without temporal immersion. Orders could be issued without delay, revised instantly, amplified across vast territories. The human voice gained reach without travel, impact without encounter.

The world became responsive in real time.

The telegraph introduced a new moral asymmetry: **command without contact.**

Those who decided were increasingly insulated from the consequences of their decisions. Effects occurred elsewhere. Suffering unfolded offstage. Responsibility stretched thin across wires and offices.

No one intended cruelty. But cruelty became easier.

When action is separated from presence, empathy becomes optional. When consequence is delayed or displaced, responsibility becomes abstract.

Judaism had always resisted this abstraction. Its law insists on proximity—between judge and judged, witness and event, speaker and speech. Words matter because they are spoken *to someone*. Actions matter because they are done *to someone*.

The telegraph disrupts this intimacy.

Jewish thinkers sensed this danger early.

A nineteenth-century Hasidic teaching captures it with startling precision. When asked what could be learned from new technologies, Rabbi Avraham Yaakov of (Sadagóra) in the 19th century answered:

From the train: that one moment can cost everything.

From the telegraph: that every word is counted and charged.

From the telephone: that what is said here is heard there.

This was not technophobia. It was moral perception.

Acceleration intensifies responsibility rather than dissolving it—but only if responsibility is consciously preserved. When speed increases without ethical reinforcement, words multiply faster than accountability.

And yet—even after the collapse of distance—something crucial still holds.

Humans still originate messages. Humans still decide content. Humans still interpret meaning. Machines transmit, but they do not judge. Wires carry signals, but they do not choose outcomes.

Responsibility is strained, but not yet severed.

The human being remains the site of intention.

By the end of the nineteenth century, humanity has achieved something unprecedented:

- Time can be compressed
- Space can be collapsed
- Power can be projected instantly
- Action can occur without presence

And yet, despite all this, the moral architecture inherited from earlier ages still functions—barely.

Judgment is slower than transmission, but it exists. Interpretation lags behind action, but it survives. Responsibility is stretched thin, but it is still human.

This condition will not endure indefinitely.

The train and the telegraph prepare the world for something more radical than speed or reach.

They prepare it for **automation of decision.**

Once action no longer requires presence, the next temptation is to remove deliberation altogether. Once messages move faster than thought, the next step is to let systems decide which messages matter.

The collapse of distance sets the stage for the collapse of judgment.

That collapse will not arrive through steel or wire—but through abstraction.

And when it arrives, the human monopoly on meaning will finally be broken.

Chapter 9

The Digital Turn:
When Reality Became Code

The revolutions described so far transformed the conditions of power, but they did not alter its center. Science made the world intelligible. Industry conquered time. Railways and telegraphs collapsed distance. Yet in every case, meaning, judgment, and responsibility still originated in the human mind.

The Digital Revolution changes this condition at its root.

It does not merely accelerate action or expand reach. It **abstracts reality itself.** For the first time, the world is no longer encountered primarily as matter, place, or event, but as information. Reality becomes legible not only to humans, but to machines. When reality becomes code, meaning detaches from presence. Responsibility loses its last anchor in the human body.

This shift marks a quiet but decisive threshold: **meaning detaches from embodiment.**

The digital turn begins when information is separated from its physical carrier. Language no longer requires paper. Images no longer require objects. Memory no longer requires bodies or places. Everything that once depended on material presence can now be encoded, copied, transmitted, and stored as data.

This is not a technical footnote. It is an ontological transformation.

When reality becomes code, it becomes manipulable in ways that physical reality never was. Code can be duplicated without loss, transmitted without delay, recombined without cost. It can be processed at speeds that exceed human cognition.

The world does not simply move faster. It becomes **processable.**

Abstraction has always been a source of power. Writing abstracts speech. Money abstracts value. Law abstracts justice. Each abstraction allows control at scale—but each also introduces moral risk.

The digital abstraction is unprecedented because it is total.

Everything—speech, desire, movement, preference, attention—can be translated into data. Once translated, it can be analyzed, predicted, and optimized. Reality no longer resists interpretation. It yields.

This produces a new form of authority: authority grounded not in force or presence, but in pattern.

Earlier technologies were instruments. They extended human faculties but remained passive. They waited for instruction. They did not initiate.

Digital systems behave differently.

They sort, rank, recommend, filter, and prioritize. They shape what is seen and what is ignored. They mediate access to reality itself. Increasingly, they operate continuously, autonomously, and at scale.

Yet even here, something still appears intact.

Humans design systems. Humans set objectives. Humans remain "in the loop." Judgment, it seems, has merely been assisted—not replaced.

This appearance is deceptive.

Digital systems create an illusion of mastery because they are built by humans. But once deployed, they operate in environments too complex and dynamic for human supervision to remain meaningful.

No individual understands the total system. No single decision explains the outcome. Responsibility disperses across architectures, updates, feedback loops, and emergent behaviors.

When something goes wrong, there is no clear author—only processes.

This diffusion does not eliminate responsibility. It obscures it.

Judaism has always warned against precisely this condition: action without a face, power without a name, authority without an address.

Despite this pressure, one boundary still holds at the end of the digital turn.

Machines process information. They do not interpret meaning.

Algorithms sort data, but they do not understand what matters. Systems optimize outcomes, but they do not know why those outcomes should matter. Context, value, and moral significance still require human judgment.

This is the last human monopoly.

And it is fragile.

Once reality is fully abstracted—once language, images, memory, and choice are encoded—the next step becomes imaginable.

If patterns can be processed, why not predicted?

If predictions can be generated, why not acted upon?

If action can be automated, why retain human deliberation at all?

The digital turn does not yet answer these questions. But it makes them unavoidable.

For the first time, the human being is no longer the sole site where reality is made sense of. Interpretation itself begins to migrate into systems. Judgment is no longer merely assisted—it is prepared for simulation.

This is the moment at which the story changes.

Up to this point, every transformation in human history—however disruptive—has preserved a single, decisive assumption: **that meaning originates in the human being.**

Science made the world intelligible, but humans interpreted its significance. Industry multiplied power, but humans decided how it would be used. Railways and telegraphs collapsed distance, but humans still chose what to say and when to act.

Digital systems abstracted reality, but humans remained the final arbiters of value, context, and judgment.

That assumption now stands exposed.

What we have been calling "progress" has not simply increased power. It has steadily **thinned the distance between intention and execution**, eroded the pauses in which judgment forms, and displaced responsibility into systems too complex to be fully seen. At each stage, something human remained at the center—until now.

The danger ahead is not that machines will become powerful. Power has always been dangerous. The danger is that **judgment itself may cease to require a human subject.** When interpretation becomes automated, when explanation is generated without understanding, when decisions are executed faster than deliberation can occur, responsibility no longer disappears—it becomes unlocatable.

Judaism has always insisted that responsibility must have an address. Someone must be able to answer. Someone must be able to say *hineni*—here I am.

A civilization in which no one can say that has not solved the problem of power. It has evaded it.

For the first time, humanity is building systems that do not merely act **for** us, but begin to act **instead of** us—systems that learn, infer, recommend, and soon decide.

This is not the arrival of a new tool. It is the arrival of a new participant.

The chapters that follow will examine how non-biological intelligence emerges, why it cannot be understood as a continuation of previous technologies, and what happens when judgment itself becomes scalable.

The question is no longer whether power can be controlled.

It is whether **responsibility can survive when intelligence is no longer human by default.**

The last human monopoly is about to be tested.

What follows is not an extension of earlier technologies, but a rupture in the moral economy itself.

Chapter 10

The Emergence of Non-Biological Intelligence

U ntil now, every civilization-shaping invention—writing, printing, machines, computation—depended on a single condition: **human interpretation remained the final site of meaning.** Tools extended capacity. Systems increased scale. But initiation, judgment, and responsibility stayed anchored in human minds.

That condition no longer holds.

Artificial intelligence marks the first moment in history when intelligence itself begins to detach from biology—not metaphorically, but functionally. This is not a new instrument of power. It is a new form of agency entering the human world.

Previous technologies obeyed. They waited. They executed instructions. Even the most complex machines remained passive until activated by human intent.

AI systems are different.

They do not merely store information or accelerate calculation. They model patterns of reasoning. They infer. They generate. They anticipate. They interact with reality in ways that resemble understanding—without possessing it.

For the first time, the act of *making sense of the world* is no longer exclusively human.

This is the civilizational rupture.

AI is a set of human-built systems designed to interpret inputs, generate responses, and act in the world without requiring constant human instruction.

It does not think as humans think. It does not know as humans know. It does not understand meaning, intention, or value. What it does is detect patterns at scale and operate upon them with speed and consistency no human can match.

And yet—because language, judgment, and choice themselves exhibit patterns—AI can simulate the surface of cognition with uncanny fidelity.

The danger is not confusion about what AI *is*. The danger is confusion about what it *replaces*.

Early pioneers of artificial intelligence believed intelligence could be engineered directly from formal logic. If reasoning followed rules, those rules could be coded. If knowledge could be represented symbolically, machines could manipulate it.

This approach failed repeatedly.

Decades of effort produced systems that worked in narrow domains but collapsed in open environments. Human intelligence proved too contextual, too embodied, too emotionally saturated to be reduced to formal schemes.

A second approach emerged—not to replicate intelligence directly, but to **approximate it statistically.** Instead of programming understanding, engineers trained systems on massive amounts of human behavior: language, images, preferences, decisions.

AI did not learn what things *mean*. It learned how humans *behave* when they appear to mean something. This shift changed everything.

Modern AI systems are powerful precisely because they are not intelligent in the human sense.

They do not hesitate. They do not doubt. They do not tire. They do not carry memory as burden or trauma. They do not inherit culture as responsibility. They do not answer to history.

They operate without inner life.

And yet, because they can generate language, explanation, and recommendation, they increasingly occupy roles once reserved for human judgment: advisor, translator, evaluator, even creative partner. This is not imitation at the margins. It is intrusion at the core.

From the beginning, AI has carried a promise—and a threat—of self-acceleration.

If machines can assist human thinking, why not let them design better machines? If intelligence can be improved iteratively, why assume a ceiling at the human level?

This idea—first articulated in the mid-twentieth century and later popularized as the "Singularity"—rests on a simple premise: non-biological intelligence can combine speed, memory, and scalability in ways biological intelligence cannot. Whether or not such an explosion occurs is beside the point.

The ethical rupture happens much earlier—**the moment we allow systems to decide without moral delay.**

Some argue that AI is best understood not as artificial intelligence at all, but as a new form of social collaboration: humans and machines thinking together.

This is partially true—and deeply misleading.

Collaboration presupposes symmetry of responsibility. It assumes that all participants can be held to account. But AI systems cannot answer. They cannot explain themselves morally. They cannot stand behind their outputs. They can only generate more output.

When collaboration lacks reciprocity, it quietly becomes substitution.

The decisive question is not whether machines can think. It is whether

thinking is what makes us human.

If intelligence is treated as the measure of worth, then machines will inevitably surpass us. If speed, optimization, and predictive accuracy become the highest virtues, human judgment will appear slow, biased, and expendable.

Judaism offers a different measure.

The human being is not irreplaceable because he thinks. He is irreplaceable because he is answerable.

AI has no covenant. No commandedness. No *hineni*. It cannot say "here I am." It cannot take responsibility for consequences that extend beyond its function.

This is not a technical limitation. It is an ontological one.

With AI, humanity crosses a line it has never crossed before.

For the first time, intelligence enters history not as nature or fate, but as an engineered presence—one that speaks, judges, and acts within human systems without sharing human responsibility.

This is why AI cannot be understood as the next technology.

It is the first non-human presence to enter the human space of speech, judgment, and decision—the very space in which Judaism locates dignity.

What follows will examine what happens when that space is no longer exclusively human. The question is no longer whether machines will become intelligent. It is whether human beings will remain responsible.

The public debate around artificial intelligence is dominated by a single question: *Can machines think?* This question is not only misleading—it is dangerous. It directs attention to the wrong axis of the problem and allows the real moral rupture to proceed unnoticed.

Judaism would recognize this immediately. Civilizations collapse not because they misunderstand capacity, but because they misname responsibility.

To frame AI as a question of intelligence is to accept a modern mistake: that intelligence is what makes human beings human. This assumption is recent, culturally specific, and deeply unstable.

For most of human history, intelligence was never treated as the source of dignity. Children, the elderly, the ill, and the unlearned were not considered less human because they knew less or reasoned more slowly. What bound human beings together was not cognitive parity, but moral standing.

Judaism never defined the human being as *the one who thinks best*. It defined the human being as *the one who can be addressed and held to account*.

Once intelligence becomes the measure of worth, comparison becomes inevitable. Ranking follows. Replacement becomes imaginable. If intelligence is what matters, then superior intelligence will always justify dominance.

This is not a future danger. It is an old one, returning in digital form.

Artificial intelligence is not a new intelligence competing with ours. It is a different category altogether.

Human intelligence is inseparable from:

- embodiment
- vulnerability
- mortality
- memory as burden
- time as limitation
- relationship as obligation

AI intelligence is none of these things. It is statistical pattern-matching operating at scale. It has no inner life, no stake in outcomes, no exposure to consequence.

To compare these two under the single label "intelligence" is to commit a category error—like comparing law to gravity or prayer to electricity.

Judaism insists that meaning emerges not from cognition alone, but from *answerability*. Intelligence without answerability is not wisdom; it is power without restraint.

The true question is not whether machines can think. It is whether **decision, judgment, and authority can be exercised without a responsible subject.**

That question has nothing to do with IQ, creativity, or problem-solving ability. It has everything to do with moral structure.

A system can outperform humans at every cognitive task and still be morally irrelevant—*unless* we allow it to decide, recommend, prioritize, or act in ways that shape human lives.

The moment we delegate judgment rather than execution, intelligence ceases to be the issue. Responsibility becomes the issue.

The obsession with intelligence is not accidental. It is psychologically reassuring.

If AI is framed as a contest of minds, humans can comfort themselves with benchmarks: creativity tests, consciousness debates, emotional simulations. As long as machines "aren't really intelligent," the moral crisis feels postponed.

But the crisis does not wait for machines to become conscious. It begins the moment:

- decisions are automated
- explanations are generated without understanding
- outcomes occur without anyone able to say "I chose this"

Judaism would call this a collapse of addressability.

Intelligence becomes dangerous only when it is detached from covenant.

Covenant means obligation that cannot be optimized away. It means being bound to the other even when efficiency argues otherwise. It means standing inside history rather than above it.

AI has no covenant. It cannot be commanded. It cannot be shamed. It cannot repent. It cannot say *hineni*—here I am.

This is not a technical limitation to be solved. It is a boundary that defines the moral universe.

To place intelligence without covenant in positions of judgment is not progress. It is abdication.

The correct framing is not: *Can machines think?*

It is:

- *Who is responsible when decisions are made?*
- *Where does answerability reside?*
- *What happens when no one can be addressed?*

Judaism has survived every technological upheaval because it never confused power with legitimacy or intelligence with worth. It understood that civilization collapses not when tools grow strong, but when responsibility grows thin.

If we continue to debate AI as an intelligence problem, we will regulate performance while ignoring authority. We will audit outputs while evacuating judgment. We will marvel at capability while forgetting obligation. And by the time machines appear intelligent enough to frighten us, responsibility will already be gone.

AI must be evaluated not by how well it thinks, but by **what kind of moral world it creates.**

The question is not whether AI resembles us.

It is whether it leaves room for us to remain human.

Judaism does not ask whether power can be achieved. It asks whether power can be governed. That question has returned—sharper than ever. And it will not be answered by intelligence. It will be answered by responsibility—or by its disappearance.

Chapter 11

AI and the Re-Invention of Human Desire

A rtificial intelligence does not merely influence what human beings choose. It intervenes in how desire itself forms- and in doing so, it alters the conditions under which moral responsibility can arise at all

Judaism has never treated desire as morally neutral. Desire is the raw material of action, but responsibility depends on the capacity to recognize, examine, and sometimes resist it. A human being can answer only for what he can encounter as his own.

What changes under artificial intelligence is not simply what human beings choose, but **how preference and inclinations are formed**—often before judgment, reflection, or resistance can occur. Previous technologies expanded what human beings could do. Artificial intelligence reshapes what human beings come to want. This is a more intimate intervention, and therefore a more dangerous one.

Desire has always moved human action. It directs attention, animates choice, and gives force to intelligence. Judaism recognized this long before modern psychology. Desire is neither mere instinct nor pure freedom. It is formed—disciplined, distorted, or refined—by law, culture, memory, and moral demand.

AI systems learn from patterns of behavior: what people read, watch, buy, repeat, and linger over. From these traces, they infer preferences. But inference quickly becomes anticipation. When a system can predict what a person is likely to want, it can present that option in advance of reflection. What begins as personalization becomes persuasion. The sequence reverses: desire no longer precedes choice; choice begins to generate desire. When this happens, responsibility loses its point of entry.

This feedback loop is subtle but profound.

Human beings have always been influenced by their environments. In the past, that influence came largely from shared narratives, traditions, authorities, and lived encounters with others. Instead of being shaped by shared human judgment—teachers, texts, communities, and debate—people are increasingly steered by systems that rely on recurring patterns drawn from past human behavior, presenting what has already been done as the most likely next step, without asking whether that step is good, meaningful, or responsible.

Over time, this alters how desire itself is understood. Desire is treated as something legible in advance, reducible to patterns inferred from prior choices, and projected forward as likelihood. What matters is not why something should be wanted, but how often it has been wanted before.

Judaism has never understood desire in this way. Desire is not transparent to the self, nor immediately trustworthy. The human being does not automatically know what he wants—or why. This opacity is not a flaw to be eliminated; it is the condition under which moral growth becomes possible.

To desire rightly requires reflection, interruption, and sometimes refusal. Jewish law does not seek to extinguish desire, but to educate it—to slow it down, redirect it, and embed it within obligation. Appetite is never sovereign.

AI short-circuits this process.

By continuously presenting individuals with options that align with past behavior, AI systems remove the pause through which desire becomes conscious. The question *"Do I want this?"* quietly disappears. Desire arrives already shaped. This is not manipulation by ideology. It is manipulation by familiarity.

Moral formation requires struggle. The tension between impulse and restraint, between appetite and obligation, is where character forms. Judaism assumes this struggle is unavoidable—and productive. AI weakens it.

When systems optimize for engagement, comfort, and continuity, they minimize disruption. They do not ask what should unsettle us. They ask what will keep us moving smoothly. Over time, this produces a subtle moral atrophy. Not because people become wicked, but because they become unpracticed in resistance. Desire no longer meets law as a limit that must be reckoned with; it meets repetition that encourages it forward.

The inner life grows quieter—not deeper.

AI does not only shape desire; it reshapes imagination by narrowing what individuals come to see as possible, likely, or worth pursuing.

On the one hand, AI expands imaginative possibility. It can visualize worlds never seen, generate voices never heard, simulate histories never lived. Creativity becomes accessible at unprecedented scale. On the other hand, imagination risks becoming automated.

When images arrive without effort, when narratives assemble themselves, when creation requires no struggle, imagination shifts from an act to a consumption. The labor that once bound meaning to effort dissolves.

Judaism has always insisted that meaning requires work. Study is labor. Interpretation is struggle. **What Judaism calls holiness is not spontaneity or purity of feeling, but the disciplined** form of life that emerges when responsibility is sustained over time under concrete conditions.

AI can simulate empathy convincingly. It can mirror language of care, recognize emotional cues, and respond with calibrated reassurance. But this simulation lacks interiority.

Empathy in Judaism is not affect alone. It is responsibility for the other. To feel without obligation is sentiment. To respond without answerability is performance. AI does not bear the weight of the other's vulnerability. It does not risk itself. It cannot be commanded.

This matter because simulated care, when widespread, can displace real responsibility. If comfort is always available, the demand to act may feel less urgent.

At the heart of Judaism's moral vision is covenant: a binding relationship that shapes desire through obligation. One does not ask only "What do I want?" but "What is asked of me?" AI knows nothing of covenant. It optimizes satisfaction without reference to worth. It amplifies preference without reference to consequence. It learns desire without learning responsibility.

If AI becomes the primary mediator of human attention, recommendation, and imagination, desire itself will be trained outside any moral horizon. This is not neutrality. It is formation without aim.

The problem is not that AI thinks. The danger is that human beings may stop thinking about what they want—and why. A civilization that allows desire to be engineered without moral interruption will not collapse dramatically. It will drift. It will grow comfortable, efficient, and shallow. Responsibility will not be rejected. It will be forgotten.

Judaism has endured precisely because it refused to let desire govern itself. It insisted that the human being is not defined by appetite, but by the capacity to bind appetite to meaning. Artificial Intelligence challenges that insistence at the deepest level.

If desire is no longer formed through struggle, memory, law, and covenant, then intelligence—human or artificial—will serve nothing but itself. And a world in which desire no longer answers to responsibility is already post-human, long before any machine surpasses us.

The argument so far has not been about technology alone, but about formation: how desire, imagination, and judgment are shaped before a choice is even felt as choice. This is why the usual institutional response—regulation—cannot reach the core of the disruption. Rules can limit uses and punish harms, but they cannot restore the interior conditions under which responsibility becomes possible. To see why, we must turn from the psychology of desire to the structure of governance itself.

Chapter 12

Why Regulation is Not Enough

E very technological disruption provokes the same institutional instinct: to regulate it.

Commissions are formed. Frameworks are drafted. Guidelines are issued. Hearings are convened. The public is reassured that control is being restored and that the future will be managed through oversight, compliance, and guardrails. This reflex is not cynical. It is sincere. Regulation is how modern societies express responsibility when confronted with powers they did not anticipate.

But sincerity does not equal sufficiency.

Regulation feels like action because it is visible and familiar. It speaks the language institutions know how to speak: risk assessment, liability, enforcement, compliance. It promises containment. It offers governance without requiring a rethinking of foundations. This is precisely why it fails at moments of civilizational rupture—when inherited assumptions about reality, authority, and responsibility no longer hold.

Regulation works when the problem is misuse or excess within an otherwise stable system. It presumes that technologies remain tools—bounded, instrumental, and subordinate to human intention. Artificial intelligence does not fit this model. It is not merely risky. It is reshaping the conditions under which choice, judgment, and responsibility arise in the first place.

Regulation assumes that decisions are made by identifiable agents who can be addressed, questioned, and held to account. It presumes that intention precedes action and that causality can be traced. Artificial intelligence begins to dissolve these assumptions. This is why regulation often arrives quickly—and reveals its limits just as quickly.

Modern institutions are designed to govern outcomes, not formation. They respond to harms after they occur; they do not shape the conditions under which harms become likely or invisible. Regulation can prohibit certain uses of AI, require disclosures, and impose limits—but it cannot answer the deeper question: *what kind of human being is being formed in an environment where judgment, recommendation, and decision are increasingly automated?*

This is not an administrative gap. It is a conceptual one. Much of what passes for AI ethics today consists of regulatory language applied to moral questions: transparency, accountability, fairness, explainability. These are not trivial concerns. But they remain procedural. They ask whether systems meet standards. They do not ask whether delegating judgment to systems is itself legitimate.

The distinction is decisive. Regulation governs how tools are used. Responsibility governs whether certain uses should exist at all. Judaism has always drawn this line sharply. Not everything that can be done may be done. Not every power that can be exercised should be exercised. These are not regulatory insights. They are moral ones.

Law operates retrospectively. Harm occurs; patterns are identified; rules are drafted; enforcement follows. Artificial intelligence does not repeat the past—it recombines it. Learning systems adapt continuously. By the time a rule is written, the system it targets has already changed. Regulation is therefore not useless, but it is always late.

This is not a defect of law. It is its moral condition. Law is slow by design. It requires deliberation, debate, interpretation, and consent. Delay is how law remembers. Pause is how judgment becomes possible.

Artificial intelligence operates according to the opposite logic. It acts continuously. It recommends, updates, and adapts without stopping. When decisions occur faster than deliberation can take place, hesitation comes to appear as failure. Reflection looks inefficient. Judgment gives way to momentum.

Judaism has always understood that judgment requires interruption. Its laws, rituals, and rhythms exist not to constrain life arbitrarily, but to slow it—to create spaces in which responsibility can emerge. A civilization that surrenders the power to pause surrenders the conditions under which responsibility can arise at all.

This exposes the deepest limit of regulation. Rules presuppose a responsible subject—someone who can be addressed, questioned, and who can answer. Artificial intelligence erodes this address. Designers point to users. Users point to systems. Institutions point to compliance. No one can truthfully say: *I chose this.*

Judaism insists that responsibility must have an address. Someone must be able to say *hineni*—here I am. A system that produces effects without such an address is not merely difficult to regulate. It is morally incoherent.

Regulation can slow harms. It cannot restore responsibility once judgment has been displaced. If a civilization allows systems to set the tempo of decision-making, responsibility does not disappear. It becomes ceremonial. Rules remain. Oversight remains. But judgment no longer governs action.

For this reason, the response to artificial intelligence cannot begin with policy. It must begin with anthropology—with a renewed insistence on what a human being is, and on what cannot be delegated without loss. The question is no longer whether institutions will adapt.

It is whether responsibility itself will still have a home.

Chapter 13

AI and the End of Moral Responsibility

M oral responsibility is not instantaneous. It does not arise at the moment of stimulus, nor at the speed of reflex. Responsibility presupposes delay: a pause between what happens and how one responds. In that interval—often narrow, sometimes agonizing—judgment becomes possible. Without delay, there is action, but not responsibility; behavior, but not moral agency.

What ends here is not morality itself, but the conditions under which responsibility can still be carried.

Human moral life has always depended on this interval. The biblical command *"Hear, O Israel"* does not instruct immediate compliance; it demands attention before action. Rabbinic law institutionalizes delay through deliberation, debate, dissent, and procedure. Courts postpone judgment. Witnesses are cross-examined. Even God, in the biblical imagination, is portrayed as pausing—regretting, reconsidering, relenting. Moral meaning emerges not from speed, but from restraint.

Delay is not inefficiency. It is the moral condition of possibility.

Artificial intelligence, by contrast, is built to eliminate delay. Its promise lies precisely in compression: faster decisions, real-time optimization, continuous response. Where human judgment hesitates, AI executes. Where moral reasoning struggles, AI calculates. This is not an accidental feature of the technology; it is its defining aspiration.

Yet the removal of delay does not merely accelerate action. It erodes responsibility.

Responsibility requires more than choosing among options. It requires being *answerable* for a choice—capable of explaining why this action was taken rather than another, under conditions that could have allowed for a different response. Answerability presumes time: time to reflect, to weigh, to resist impulse or pressure. When systems act faster than reflection can occur, responsibility is not transferred; it is dissolved.

This is why the contemporary language of "human-in-the-loop" is inadequate. Placing a human somewhere in a high-velocity system does not restore moral agency if the system's tempo renders meaningful intervention impossible. A human reduced to approving outputs at machine speed is not exercising judgment; he is ratifying inevitability.

The problem is not that machines make decisions. The problem is that they make decisions at a speed that precludes moral interruption.

Historically, moral systems evolved precisely to slow things down. Rituals, laws, councils, sabbaths, appeals—all are technologies of delay. They interrupt momentum. They create space where power must wait for justification. Even warfare, at its most regulated, was constrained by declarations, chains of command, and rules of engagement designed to preserve some minimal interval between capability and use.

AI collapses these intervals. It produces environments in which action is continuous, adaptive, and self-updating. Decisions are no longer discrete events but ongoing processes. There is no clear "moment" of choice—only streams of output responding to streams of input. In such a setting, asking *"Who decided?"* becomes increasingly incoherent.

And where decision cannot be located, responsibility cannot be assigned.

This is the first rupture: moral responsibility requires delay, but AI governance is structured around its elimination. Not because of malice, but because efficiency has become the supreme value of the system. What disappears is not ethics as a set of rules, but responsibility as a lived human capacity.

The danger, then, is not that AI will act immorally. It is that it will act *amoral*—outside the temporal conditions that make moral judgment possible at all.

What follows from this is not merely a regulatory challenge, but a civilizational one. If delay is the cradle of responsibility, a world without delay is a world in which responsibility has no place to stand.

The next question is unavoidable: if responsibility requires delay, what happens to moral life when speed becomes sovereign?

Chapter 14

Speed, Sovereignty, and the Disappearance of Judgment

M odern power no longer announces itself through authority, law, or command. It announces itself through speed.

Judgment requires the power to interrupt action. Under speed, that power disappears.

What governs contemporary systems is not who decides, but how fast decisions are executed. Speed has become sovereign—not as an ideology explicitly defended, but as an operational fact no one feels authorized to interrupt. In such a world, judgment does not fail dramatically; it quietly becomes irrelevant.

Judgment presupposes a subject who can be addressed, questioned, and held to account. It presupposes a moment in which action could have been otherwise. Speed dissolves that moment. When systems operate continuously and adaptively, across many decisions and actors, there is no discrete act to judge—only momentum.

This is not a new temptation. Human beings have always been drawn to speed as a form of power. What is new is that speed has escaped human limits.

In premodern societies, speed was constrained by bodies, distances, fatigue, and friction. Even tyrannical power moved slowly. Orders took time to travel. Enforcement lagged behind intention. These delays were not merely technical constraints; they functioned as moral buffers. They allowed for reconsideration, resistance, misinterpretation, even mercy.

Modern bureaucracies reduced some of these delays, but they replaced them with procedures. Files moved faster than messengers, but still had to pass through hands. Authority remained legible because it was mediated by roles, offices, and signatures.

AI-driven systems eliminate mediation altogether. They act immediately, without pause, review, or the possibility of human interruption.

In such systems, sovereignty migrates from judgment to optimization. What matters is not whether an action is justified, but whether it improves performance metrics: efficiency, accuracy, engagement, risk reduction, profit. These metrics are not immoral in themselves. But they are indifferent to meaning.

Judgment, by contrast, always involves choosing what matters and accepting the consequences of that choice. It involves asking questions that cannot be reduced to performance: Is this fair? Is this fitting? Is this permissible? Is this ours to do?

These questions slow action down. That is precisely why they are increasingly treated as obstacles.

The result is a structural inversion: instead of speed serving judgment, judgment is asked to justify its interference with speed.

This inversion explains a defining feature of contemporary moral confusion. When harm occurs, inquiries are launched—but always after the fact. Investigations examine failures in oversight, data quality, model bias, or governance frameworks. Rarely do they address the deeper issue: the system worked as designed. It acted faster than judgment could intervene.

Responsibility is then redistributed into abstractions: "the system," "the process," "the pipeline," "the market." Each explanation is technically accurate—and morally evasive.

What disappears is not blame, but answerability.

Judgment requires someone who can say: *I acted, and I could have acted otherwise.* Speed-based systems are engineered to eliminate precisely that condition. They aim to ensure that once certain inputs are present, outputs follow necessarily. Optionality is treated as noise. Hesitation is treated as inefficiency.

But necessity is the enemy of responsibility.

This is why appeals to "ethical AI" often feel hollow. Ethics presumes agents who can pause, reflect, and restrain themselves. Sovereign speed produces systems that cannot stop without external disruption—and human actors who no longer experience themselves as authors of outcomes.

The deeper danger is not technological domination, but moral atrophy. As judgment becomes structurally unnecessary, human beings lose the habit of exercising it. They learn to defer—not to authority, but to tempo.

Judaism has always resisted this logic. Its most radical claim is not theological but temporal: that power must submit to time. The Sabbath is not a rest from labor alone; it is a suspension of sovereignty. Even God, in the rabbinic imagination, limits divine action through law and rhythm. Creation is structured to include interruption, restraint, and rest.

Speed is never allowed to rule.

This is not nostalgia. It is a diagnosis. A civilization that cannot interrupt itself cannot judge itself. And a civilization that cannot judge itself cannot be responsible for what it becomes.

Chapter 15

The Vanishing Subject

M oral responsibility does not float freely. It attaches to someone.

Ethics presupposes a subject who can be addressed, accused, summoned, and bound. Commands require an addressee. Judgments require a bearer. Responsibility is not a property of systems; it is a condition of persons. When the subject disappears, morality does not evolve—it evaporates.

This is the final consequence of AI-driven speed and scale: not the corruption of moral norms, but the erosion of the moral subject itself.

The subject vanishes not because consciousness disappears, but because responsibility can no longer be located.

For most of human history, action could be traced. Even when authority was diffuse, someone could still be named. A king, a general, a judge, a council, a merchant. Responsibility might be evaded, displaced, or denied—but it could still be *claimed*. The grammar of moral life assumed that somewhere, someone had acted.

Contemporary AI systems undo this grammar.

Decisions are produced by pipelines rather than persons, by models rather than minds, by probabilistic outputs rather than intentions. Responsibility is distributed across designers, trainers, deployers, users, regulators, data sources, and feedback loops—until it no longer coheres anywhere. Each participant contributes causally, but no one authors the outcome.

This is not collective responsibility in the classical sense. Collective responsibility presumes a shared subject—a people, a community, a covenant—capable of acknowledging guilt and repair. What we are witnessing instead is responsibility without a subject: causation everywhere, answerability nowhere.

The question *"Who is responsible?"* increasingly receives answers that are structurally correct and morally useless.

- The engineer followed specifications.
- The company followed market incentives.
- The regulator followed existing law.
- The system optimized as designed.

Each statement is true. Together, they form an alibi without an agent. The moral subject vanishes not through denial, but through fragmentation.

This fragmentation is often mistaken for sophistication. We speak of "socio-technical systems," "emergent behavior," "complex causality." These descriptions are accurate—but they obscure a deeper loss. Complexity becomes a substitute for accountability. Explanation replaces responsibility.

Yet moral life cannot survive explanation alone.

To be a subject is not merely to cause effects. It is to be *addressable*—to stand in a relation where one can be commanded, questioned, and held. The biblical *"Where are you?"* is not a request for location; it is the birth of responsibility. It assumes someone who can answer.

AI systems cannot answer in this sense. They can generate responses, but they cannot stand behind them. They cannot say "I should not have done this." They cannot repent, repair, or restrain themselves. They cannot bear guilt. And they cannot assume obligation beyond optimization.

More troubling is what happens to humans embedded within such systems. As agency is absorbed into processes, people cease to experience themselves as responsible subjects. They become operators, supervisors, stakeholders, users—roles defined by function rather than accountability.

This is not moral failure. It is moral displacement.

The human being is still present, but no longer positioned where responsibility can attach. One cannot answer for outcomes one never truly chose, at speeds one could not interrupt, in systems one does not control. The subject is not absolved; it is bypassed.

Judaism saw this danger long before algorithms. It understood that responsibility is fragile—and that power, left unchecked, seeks to dissolve it. This is why it insists so relentlessly on naming subjects: *you shall, you shall not, you are responsible for your brother.* Even kings are bound. Even God, in the rabbinic imagination, submits to law.

Responsibility survives only where someone can still say: *Here I am.*

AI civilization moves in the opposite direction. It builds systems that work precisely because no one must say that. Outcomes emerge. Metrics improve. Harm is regretted. Responsibility disperses.

What is lost is not control, but addressability.

This is the end of moral responsibility in its classical sense—not because humans become evil, but because the architecture of action no longer requires a responsible self.

The final question, then, is not how to regulate AI, nor how to make it ethical. It is whether a civilization organized around systems that do not need subjects can still sustain human beings as moral agents at all.

If responsibility requires a subject, and the subject is vanishing, then the crisis we face is not technological.

It is anthropological.

Chapter 16

Responsibility Without Refuge

Responsibility becomes most visible when all refuges fail. When responsibility becomes unbearable, human beings look for refuge.

They look for something to hide behind: law, procedure, necessity, system, history. Refuge does not deny responsibility outright; it relocates it elsewhere. It offers shelter from answerability without requiring moral collapse. One can still act—often decisively—without having to stand exposed to judgment.

For most of history, moral traditions worked precisely to *deny* this refuge. They insisted that even when circumstances constrain action, responsibility does not disappear. The question was never *"Did you cause this?"* but *"How did you respond when addressed?"*

What AI introduces is not merely a new refuge, but a structurally perfect one.

Systems absorb responsibility so completely that no hiding is required. No one needs to evade accountability because no one is positioned to receive it in the first place. Refuge becomes architectural rather than psychological.

This marks a profound shift.

In earlier moral failures—bureaucratic atrocities, ideological violence, colonial domination—perpetrators still experienced themselves as agents. They justified themselves, rationalized their actions, or appealed to higher causes. Their defenses were morally corrupt, but they still presupposed responsibility. They answered badly—but they answered.

AI systems remove the need even for bad answers.

When harm occurs, explanations proliferate: data drift, emergent behavior, unintended consequences, misalignment, edge cases. Each explanation is technically coherent. None of them answers the moral question. The very form of explanation replaces responsibility with causality.

Causality is infinite. Responsibility is singular.

To take responsibility is to accept exposure. It is to stand without refuge—to say, *this passed through me*. That exposure is what makes moral life difficult, but also meaningful. Without it, action becomes weightless.

Judaism insists on this weight. Its most uncomfortable demand is not belief, ritual, or identity, but refusal of refuge. One cannot disappear into fate, power, or system. Even when God hardens Pharaoh's heart[13], Pharaoh remains responsible. Even when circumstances are crushing, the human being is still addressed.

This is not cruelty. It is dignity.

The refusal of refuge preserves the human being as a moral subject. It insists that something remains at stake in every response—even under constraint, even under threat, even under opacity.

AI civilization offers the opposite promise: relief from burden. Let the system decide. Let optimization guide outcomes. Let responsibility dissolve into processes too complex to resist. This feels humane. It feels merciful. It feels like progress.

---

13   *Exodus* 9;12; 3:21; 7:3; 10:1; 14:18; 8:15; 7:13; 8:32: 9:7. and Rashi ad loc., who explains that Pharaoh hardens his own heart repeatedly before God is said to harden it, preserving moral responsibility.

It is none of those things.

A world without refuge may be harsh—but a world without responsibility is hollow. Where no one must answer, nothing finally matters. Harm becomes regrettable but not culpable. Injustice becomes unfortunate but unowned. Evil becomes statistical.

The deepest danger, then, is not that AI will make catastrophic decisions. It is that it will make decisions *without burden*. And that human beings, relieved of responsibility, will quietly accept that relief.

Responsibility without refuge is not sustainable without courage. It requires limits on speed, on scale, on delegation. It requires preserving spaces where judgment can interrupt momentum—and where someone must still say yes or no.

The question is no longer whether we can build ethical systems.

It is whether we are willing to live without refuge again.

The final chapter must therefore ask the most unsettling question of all: not what AI will do to us—but whether we still want to remain the kind of beings to whom responsibility can belong.

Chapter 17

The Courage to Remain Human

The crisis AI poses is often framed as a question of control: how to govern powerful systems, how to align machines with human values, how to prevent catastrophic misuse. These questions matter—but they are secondary. Beneath them lies a more unsettling one: whether we still wish to inhabit the burdened condition of being human at all.

To remain human, in the moral sense, is not to outperform machines. It is not to preserve cognitive superiority, creativity, or relevance. It is to accept responsibility under conditions that do not guarantee success, certainty, or innocence. It is to live exposed to judgment—by others, by tradition, by history, by oneself.

AI tempts us to abandon this condition gently. Not through domination, but through delegation. Not through coercion, but through convenience. Each handoff promises relief: fewer errors, faster decisions, optimized outcomes. And each handoff subtly trains us to step back—to become less involved, less answerable, less necessary.

The danger is not replacement. It is abdication.

Moral responsibility is heavy. It requires living with regret, uncertainty, and irreversibility. It demands choosing without full knowledge, acting without guarantees, and standing by outcomes one did not fully intend. AI offers an escape from this weight. It promises action without exposure, outcomes without authorship, power without burden.

That promise is profoundly anti-human.

Judaism never romanticized the human condition. It did not promise purity, clarity, or moral ease. It promised obligation. The human being is addressed before being empowered. Command precedes capability. Responsibility comes before mastery. This ordering matters.

Modern technological civilization inverts it. Capability precedes responsibility. Power expands first; ethics follows later, if at all. AI accelerates this inversion to a breaking point. We now possess systems whose reach exceeds our willingness to answer for their effects.

The question, then, is not whether AI can be made ethical.

It is whether we are willing to accept limits—on speed, on delegation, on optimization—in order to preserve spaces where human judgment still matters. Limits are not failures of imagination. They are acts of moral courage.

To remain human will require saying no—not because something is forbidden, but because something is too easy. It will require preserving delay where acceleration is possible, friction where efficiency beckons, accountability where diffusion would be convenient.

This is not a call to abandon technology. It is a call to refuse disappearance.

Civilizations are remembered not by what they could do, but by what they chose not to do. The defining question of the AI age will not be whether we built intelligent machines, but whether we preserved the fragile, costly, irreplaceable capacity to answer for what happens in our name.

Responsibility has always been a burden few wished to carry. What is new is the availability of a world that functions without it.

To choose responsibility now—to insist on remaining addressable, interruptible, and accountable—is no longer default. It is an act of resistance.

And perhaps that is what it has always meant to be human.

Conclusion

The crisis posed by artificial intelligence is not, at its core, a crisis of intelligence. It is a crisis of responsibility.

Across these chapters, one claim has emerged with increasing clarity: what is at stake is not whether machines can think, but whether human beings will continue to judge. Artificial intelligence does not merely introduce new risks that require regulation. It alters the conditions under which judgment, deliberation, and answerability can occur at all. When decision-making is automated, accelerated, and optimized beyond interruption, responsibility does not fail spectacularly. It quietly loses its place.

Modern institutions respond to power through regulation because regulation is how responsibility is expressed once judgment has already been displaced. Rules, frameworks, and oversight presume a world in which agents still decide and can still be addressed. But artificial intelligence increasingly operates before that point—shaping attention, preference, tempo, and possibility in advance of reflection. Regulation arrives after formation has already taken place. It can limit damage. It cannot restore authorship.

This is why the deepest danger is not error, bias, or misuse, but normalization. When systems act continuously and decisions occur faster than deliberation can intervene, judgment is reframed as inefficiency. Pause appears irresponsible. Speed becomes sovereign—not by argument, but by default. In such a world, no one needs to deny responsibility. It simply has nowhere to land.

Judaism recognizes this condition immediately because it has always understood responsibility as prior to system, law, or outcome. Responsibility requires a subject who can be addressed, who can say *hineni*—here I am—and who can answer for what has been done. It also requires time. Delay, interruption, and restraint are not obstacles to moral life; they are its enabling conditions. A civilization that cannot pause cannot judge. A civilization that cannot judge cannot be responsible.

The challenge posed by artificial intelligence therefore cannot be resolved through better governance alone. It demands an anthropological decision: whether judgment remains a human task that cannot be delegated without loss. Not everything that can be optimized should be. Not everything that can be automated may be. These are not technical limits. They are moral ones.

If responsibility dissolves into process, explanation, or performance, law will persist as ritual and ethics as vocabulary—but judgment will no longer govern action. What will remain is a functioning world without an addressable subject.

This book has argued that Judaism endures because it refused that outcome. It bound intelligence to obligation, power to restraint, and desire to meaning. The question now is not whether artificial intelligence will advance. It will. The question is whether responsibility will still have a home when it does.

That question cannot be answered by systems.

It can only be answered by human beings who are still willing to pause, to judge, and to answer.

Epilogue

Here I Am

T he question this book leaves us with is not what artificial intelligence will become. It is what *we* will become in its presence.

Civilizations do not fail when they lose power. They fail when they lose the capacity to answer for what they do with it. What is unprecedented about our moment is not that we are building powerful systems, but that we are building systems that function without requiring anyone to stand behind them. Action continues. Outcomes multiply. Responsibility thins.

This book has argued that moral responsibility is not a principle we apply afterward, nor a constraint we bolt onto power once it is already in motion. Responsibility is a form of presence. It requires delay, interruption, judgment, and above all, a subject who can be addressed.

The danger of the AI age is not that machines will replace human beings. It is that human beings will quietly accept being replaced as *subjects*. We will continue to act, to authorize, to benefit, to regret—but no longer to answer. We will speak in the language of systems, processes, and inevitabilities. And in doing so, human beings will lose the shared capacity to judge how to act, to recognize responsibility, and to speak meaningfully about obligation and consequence.

Judaism has always understood that the human being is not defined by intelligence, creativity, or mastery. It is defined by response. The first moral moment in the biblical narrative is not creation, but address:

"Where are you?" And the human answer—*"Hineni ("Here I am")*—is not information. It is exposure.

To say *Here I am* is to refuse refuge. It is to step forward without guarantees. It is to accept that something passes through me, and that I cannot fully disown it. This stance has never been comfortable. It has never been efficient. But it is the condition under which moral meaning exists at all.

AI tempts us to abandon this stance gently. It offers us a world that works—often better than before—without requiring us to remain visible within it. It promises relief from burden, friction, and doubt. What it cannot offer is dignity.

Dignity does not come from control. It comes from answerability.

The future will not ask whether our systems were intelligent enough. It will ask whether anyone was still there to answer when things went wrong—or when they went right. A civilization that cannot say *we did this* cannot say *we are responsible*. And a civilization that cannot say that cannot repair itself.

This book has not offered solutions, frameworks, or assurances. That is not an omission. It is fidelity to the problem. Responsibility cannot be solved. It can only be assumed—or refused.

What remains, then, is a choice. Not a technological one, but a human one. Whether we will preserve spaces where delay is possible, where judgment can interrupt momentum, where someone must still say yes or no. Whether we will accept limits—not because we are incapable, but because we are responsible.

To remain human in the age of artificial intelligence will require courage. Not the courage to build, but the courage to stand exposed. The courage to remain addressable. The courage to answer.

When the systems no longer need us, the question will still be asked.

Where are you?

And the future of moral life depends on whether there are still human beings capable of replying:

Here I am.

Appendix A

Judaism is not identical with Orthodoxy

One of the most damaging claims in contemporary Jewish life is not theological but political: the assertion—sometimes explicit, often implied—that Judaism is identical with Orthodoxy. That to be Jewish in any serious sense is to be bound by halakhah as interpreted and enforced by rabbinic authority.

This claim is not merely false. It is morally corrosive.

Judaism long predates Orthodoxy. It predates codified halakhah. It predates centralized rabbinic authority. The Bible itself is not a halakhic text. It is an argumentative one. It preserves dissent, protest, and unresolved tension. It does not reduce Jewish life to obedience. It binds Judaism to responsibility, not to institutional compliance.

Orthodoxy is right that halakhah matters. The difficulty arises when halakhah is treated as though it exhausts Judaism.

Halakhah is a legal system. Judaism is a moral civilization.

Law is indispensable. But law is not identical with responsibility. When law replaces judgment, when authority replaces address, when obedience replaces answerability, Judaism's core moral insight is betrayed in the name of its preservation.

Orthodoxy claims continuity, but often achieves **moral closure**. It treats precedent as inevitability. It sacralizes inherited authority. It discourages delay where judgment would threaten hierarchy. It replaces responsibility with compliance and calls this fidelity.

This is precisely the pattern the book has traced in modern systems:

authority disperses beyond address;

action accelerates beyond interruption;

identity hardens where judgment once oriented response.

Orthodoxy does not merely *risk* these failures. Structurally, it reproduces them.

When rabbinic authority becomes final rather than accountable, when dissent is treated as deviation, when moral questioning is framed as rebellion, Judaism's prophetic core is silenced in the name of law. Silence is demanded as humility. In truth, it t allows consequences to proceed unchecked.

Judaism never authorized such closure.

The prophets challenged priests. Kings were judged. Law itself was argued with. Even God was addressed, questioned, resisted. The tradition did not fear fragmentation because it trusted responsibility more than control.

Orthodoxy fears fragmentation because it mistrusts responsibility.

This does not mean Orthodoxy is illegitimate. It means it is **partial**—and dangerous when it claims totality. Judaism is larger than any one legal regime. It is sustained not by obedience alone, but by moral courage, dissent, and answerability.

To insist otherwise is not piety. It is power disguising itself as tradition.

Appendix B

Power, Silence, and Responsibility After October 7

October 7, 2023 did not only expose the vulnerability of Israel. It exposed the moral disorientation of Jewish power.

For centuries, Jews lived without sovereignty. Power was something exercised *over* them. Responsibility was articulated in conditions of weakness, exile, and exposure. Judaism's moral grammar—its suspicion of inevitability, its insistence on answerability—was forged in that crucible.

The return of Jewish power was never morally neutral. It demanded a transformation of responsibility, not its suspension.

What October 7 revealed is that this transformation remains unfinished.

In Israel, power has increasingly been treated as necessity. Security is invoked as inevitability. Speed replaces judgment. Military, political, and bureaucratic systems operate at a tempo that resists interruption. When harm occurs, explanation proliferates—intelligence failures, procedural breakdowns, strategic miscalculations—but responsibility disperses.

Who answers?

In the diaspora, a parallel failure appears. Jewish institutions rally reflexively around power rather than interrogating it. Silence is justified as unity. Questioning is framed as betrayal. Moral delay is treated as weakness. The language of survival overrides the language of judgment.

This is not strength. It is fear.

Judaism never taught that power excuses responsibility. On the contrary: power *intensifies* it. "Will the Judge of all the earth not do justice?" is not suspended by sovereignty. It is sharpened by it.

October 7 exposed how easily Jewish power adopts the very logics Judaism was meant to resist: inevitability, speed, unanswerable authority, bureaucratic insulation. The danger is not military failure alone. It is moral atrophy.

Power without addressability produces silence. Silence produces complicity.

This is visible in Israel's internal discourse, where dissent is increasingly delegitimized, and in global Jewish communities, where moral questions are deferred indefinitely "until after the crisis." Crisis, however, is no longer temporary. It has become permanent justification.

Judaism cannot survive this posture.

A people that once taught the world to argue with power cannot now demand exemption from judgment because power has finally arrived. Sovereignty is not redemption. It is responsibility without refuge.

The future of Jewish moral life—both in Israel and beyond—depends on whether Jews are willing to recover the courage to interrupt themselves. To reintroduce delay where speed rules. To speak where silence is rewarded. To answer where authority would prefer compliance.

October 7 was not only an attack. It was an address.

Whether Judaism will answer remains an open question.

Bibliography

Agus, Jacob B.: *The Evolution of Jewish Thought.* USA. Abelard Schuman. 1959.

Agus Jacob B.: *The Vision and the Way. An Interpretation of Jewish Ethics.* N.Y. Frederick Ungar Publishing Co. (2nd. 1969. 1st. 1966)

Amsel, Avrohom: *Rational Irrational Man. Torah Psychology.* New York. Feldheim Publishers. 1976.

Arendt, Hannah. *Responsibility and Judgment.* New York: Schocken Books, 2003.

Arendt, Hannah. *The Human Condition.* Chicago: University of Chicago Press, 1958.

Aron, Raymond. *Peace and War.* Garden City, NY: Doubleday, 1966.

Assmann, Jan. *Cultural Memory and Early Civilization.* Cambridge: Cambridge University Press, 2011.

Avishai, Bernard: *The Tragedy of Zionism: How Its Revolutionary Past Haunts Israeli Democracy.* New York. Helios Press. 2002.

Berlin, Isaiah. *Four Essays on Liberty.* Oxford: Oxford University Press, 1969.

Berman, Ari: "Innovation meets Ethics. Moral Responsibility in the Age of AI," "The Benjamin and Rose Berger Torah To-Go," "Rabbi Isaac Elchanan Theological Seminary," December

2024-Kislev 5785

Buber, Martin. *I and Thou.* New York: Scribner, 1958.

Cohen, Sagi. "How to Influence ChatGPT: Tech Startups Are Racing to Change AI's Conversation," "Haaretz," Aug 28, 2025

Ellul, Jacques. *The Technological Society.* New York: Vintage Books, 1964.

Fackenheim, Emil L. *God's Presence in History.* New York: New York University Press, 1970.

Fixler, Joshua. "Ask the Rabb-AI," September 26, 2023

Floridi, Luciano. *The Ethics of Information.* Oxford: Oxford University Press, 2013.

Freedman, David, H.: *Brainmakers: How Scientists Are Moving Beyond Computers to Create a Rival to the Human Brain.* Simon & Schuster, 1994.

Friedman, Dan. "Where is AI taking us: The path to Eden, or the road to Armageddon?" "The Jewish News of Northern California," March 7, 2025

Friedman, Dan. "As AI charges ahead, Jewish thinkers are falling behind," "The Jewish News of Northern California," March 21, 2025

Gonzalez Arocha, Jorge. "The Philosophical Misdiagnosis of AI by Yuval Noah Harari," "DIALEKTIKA," March 1, 2024

Gordis, Robert: *Judaic Ethics for A Lawless World.* New York. The Jewish Theological Seminary of America, 1986.

Gorenberg, Gershom: *The Accidental Empire: Israel and the Birth of the Settlements, 1967-1977.* New York. Henry Holt and Company. 2006. 454 pages.

Greenberg, Irving. *For the Sake of Heaven and Earth.* Philadelphia:

Jewish Publication Society, 2004.

Harari, Yuval. *21 Lessons for the 21ˢᵗ Century*. New York. Random House. 2019.

Harari, Yuval Noah. "Yuval Noah Harari argues that AI has hacked the operating system of human civilization." "The Economist," Apr. 28th, 2023.

Harari, Yuval Noah. 'Sapiens' Author Yuval Noah Harari on the Promise and Peril of AI," "The Wall Street Journal," June 19, 2025

Heilman, Samuel. *Defenders of the Faith: inside ultra-Orthodox Jewry*. New York. Schocken Books. 1992.

Heschel, Abraham Joshua. *God in Search of Man*. New York: Farrar, Straus and Giroux, 1955.

Heschel, Abraham Joshua. *The Prophets*. New York: Harper & Row, 1962.

The Hebrew Bible (Tanakh). Various editions and translations.

Jaspers, Karl. *The Origin and Goal of History*. New Haven: Yale University Press, 1953.

Jonas, Hans. *The Imperative of Responsibility*. Chicago: University of Chicago Press, 1984.

Jonas, Hans. *Technology and Responsibility*. Hanover, NH: Brandeis University Press, 1984.

Jorisch, Avi: "With AI technology rapidly advancing, ethics must evolve as well,' "The Jerusalem Post," November 2, 2025

Jubak, Jim. *In the Image of the Brain. Breaking the Barrier Between the Human Mind and Intelligent Machines*. U.S.A., Little, Brown and Company. 1992.

Kierkegaard, Søren. *Fear and Trembling*. London: Penguin Classics, 1985.

Kissinger, Henry, A; Schmidt, Eric and Huttenlocher, Daniel With

Schouten, Schuyler. *The Age of AI: And Our Human Future.* New York. Little, Brown and Company. 2021.

Koppel, Moshe. "What Artificial Intelligence Has in Store for Judaism," "Mosaic," March 4, 2024

Korol, Shayna. "What is the worst-case scenario for AI? California lawmakers want to know," "Vox," September 12, 2025

Kuhn, Thomas, S. *The Structure of Scientific Revolutions.* Chicago. The University of Chicago Press. 1996.

Kurzweil, Ray: *The Singularity is Near. When Humans Transcend Biology.* London. Penguin Books. 2006.

Lanier, Jaron: "There Is No A.I.," "The New Yorker," April 20, 2023

Ledoux, Joseph. *Synaptic Self: How Our Brains Become Who We Are.* New York. Penguin Books. 2002.

Levinas, Emmanuel. *Otherwise Than Being.* Pittsburgh: Duquesne University Press, 1981.

Levinas, Emmanuel. *Nine Talmudic Readings.* Bloomington: Indiana University Press, 1990.

Levinas, Emmanuel. *Totality and Infinity.* Pittsburgh: Duquesne University Press, 1969.

MacIntyre, Alasdair. *After Virtue.* Notre Dame, IN: University of Notre Dame Press, 1981.

Maimonides (Rambam). *The Guide of the Perplexed.* Chicago: University of Chicago Press, 1963.

Maimonides (Rambam). *Mishneh Torah.* Jerusalem: Mossad Harav Kook, 1973.

The Mishnah. Oxford: Oxford University Press, 1933.

Mitchell, Melanie. *Artificial Intelligence: A Guide for Thinking Humans.* New York. Farrar, Straus and Giroux. 2019.

Nahmanides (Ramban). *Commentary on the Torah.* Jerusalem: Mossad Harav Kook, 1960.

Novak, David. *The Jewish Social Contract: An Essay in Political Theology.* Princeton and Oxford. Princeton University Press. 2005.

Nussbaum, Martha, C. *The Therapy of Desire: Theory and Practice in Hellenistic Ethics.* Princeton University Press. Princeton, New Jersey. 1996

Ortega Y Gasset, Jose: *Man and People.* New York. W.W. Norton & Co. 1963.

Ortega Y Gasser, Jose: *Man and Crisis.* New York. W.W. Norton & Co. 1962. 217 pages.

Ortega Y Gasset, Jose: *History as a System and other Essays toward a Philosophy of History.* New York. The Norton Library.1961 [first 1941] 269 pages.

Ravitzky, Aviezer. *Messianism, Zionism, and Jewish Religious Radicalism.* Chicago. The University of Chicago Press. 1993.

Ricoeur, Paul. *Memory, History, Forgetting.* Chicago: University of Chicago Press, 2004.

Rosa, Hartmut. *Social Acceleration: A New Theory of Modernity.* New York: Columbia University Press, 2013.

Rosenzweig, Franz. *The Star of Redemption.* Madison: University of Wisconsin Press, 2005.

Sacks, Jonathan. *The Dignity of Difference.* London: Continuum, 2002.

Senor, Dan And Singer, Saul. *Start-Up Nation: The Story of Israel's Economic Miracle.* New York. Twelve. 2009.

Soloveitchik, Joseph B. *Halakhic Man.* Philadelphia: Jewish Publication Society, 1983.

Soloveitchik, Joseph B. *The Lonely Man of Faith.* New York: Doubleday, 1965.

Sprinzak, Ehud: *Brother Against Brother: Violence and Extremism in*

Israeli Politics from Altalena to the Rabin Assassination. New York: The Free Press. 1999.

The Babylonian Talmud. Vilna: Romm Publishing House, 1886–1895.

Taylor, Charles. *A Secular Age.* Cambridge, MA: Harvard University Press, 2007.

Walzer, Michael. *Exodus and Revolution.* New York: Basic Books, 1985.

Walzer, Michael. *Just and Unjust Wars.* New York: Basic Books, 1977.

Winner, Langdon. *The Whale and the Reactor.* Chicago: University of Chicago Press, 1986.

Yerushalmi, Yosef Hayim. *Zakhor: Jewish History and Jewish Memory.* Seattle: University of Washington Press, 1982.

Zuboff, Shoshana. *The Age of Surveillance Capitalism.* New York: PublicAffairs, 2019.